KILLING ME SOFTLY

KILLING ME SOFTLY

Toxic Waste, Corporate Profit, and the Struggle for Environmental Justice

Eddie J. Girdner
Jack Smith

Monthly Review Press
New York

Library of Congress Cataloging-in-Publication data
available from the publisher.

ISBN: 1-58367-083-1 (pbk)

Monthly Review Press
122 West 27th Street
New York, NY 10001

www.monthlyreview.org

Design and production by Terry J. Allen, New York, NY

Printed in Canada
10 9 8 7 6 5 4 3 2 1

Contents

Preface

In the United States, powerful multinational corporations largely design public policy and, in significant ways, plan our future. The corporate agenda rests on the necessity of profits on a global scale. Today, under the global neoliberal capitalist regime, this agenda is implemented by forcing nations to bow to International Monetary Fund (IMF) conditionality and the rules of the World Trade Organization (WTO). Under IMF conditionality, governments around the world must privatize state-owned enterprises and drastically cut social spending in order to receive loans from the IMF and World Bank merely to sustain their economies. Many countries in a debt trap to these global institutions have little choice within the current system of "global governance." Today, this actually existing form of neoimperialism forces every nation to obey the neoliberal logic of the market and accept being "governed" from Washington or marginalized in the brave new world of "globalization." Following World War II, the United States was forced to compete with the emerging centers of capitalism in Western Europe and Japan. It is not surprising, given the inherent logic of profits and competition, that environmental legislation failed to prevent environmental destruction and emerging ecological catastrophe. Today, the same awaits the global ecology, in every nation, under neoliberal global "governance."

This book is about the toxic economy created in the United States by the chemical revolution after World War II, the entirely predictable failure of legislation to limit its effects, and the many grassroots struggles of people in communities all across the United States—and the world over—who are seeking environmental justice. Today the logic of capitalism is not only "profits over people" but the corollary of that, toxic pollution over people and their health and well-being on a global scale.

We focus on a successful struggle—one that gives hope to those who may feel that the struggle is defeated before they begin, as did the people of Mercer County, Missouri, who fought against Amoco Waste-Tech from 1990 to 1993. Waste-Tech planned to site a huge hazardous waste landfill in the county, but through a number of typical environmental strategies—and "fronts"—the environmental activists of Mercer County were able to throw off the oppressor. Like many others in "sacrifice zones," they believed that they should not become the waste dump for the world.

Much of the environmental justice literature has stressed race and ethnicity as key factors in toxic imperialism. We argue, on the other hand, that while race and ethnicity are important factors, the critical cleavage nationally and on a global scale is that of class. It is largely the poor and marginalized communities lacking money and political clout who get dumped on, even as the wealthy nations and multinational corporations seek to send their wastes to the poorest countries. Indeed, as former Treasury Secretary Lawrence Summers pointed out, the logic of neoclassical economics argues for this very solution. In the United States, policies regarding the waste industry are largely purchased by corporate money through donations to politicians. Nothing illustrates this better than President George W. Bush's Texas and the policies on global warming that emerged in his first few months in the White House, shocking nations and environmentalists throughout the world.

Mercer County was targeted because the population was poor and politically marginalized—midwestern small farmers, outside of the political power structure, which is controlled by corporate America and the politicians who do their bidding at the state, national, and global levels.

The people of Mercer County won their struggle through strong and determined efforts. Today corporate power encroaches on people, whether in the Midwest, on Native American tribal lands, or in the Third World. Environmental struggles in the U.S. and around the world are a challenge to the ongoing corporate "invasion." Today there is a necessary and crucial broader and deeper movement as seen in Seattle in 1998 and Genoa in July 2001. The battles of Seattle and Genoa challenged the authority of a few rich nations and giant corporations to rule people and to destroy their way of life and livelihoods all across the globe. To challenge "global governance" by the rich nations is to radically challenge the global neoliberal system itself. It is to challenge profit over people and pollution over people. If people are to regain community empowerment, with local people controlling how their land is to be used, it is essential to defeat the tyranny of an agenda based solely on profits and capitalist accumulation.

1
The Toxics Political Economy

Scientific issues concerning emissions and potential effects on health and the environment are necessarily secondary to economic and political issues in the United States.

— John F. Chadbourne (James F. Chadbourne
Environment and Safety Programs, Inc.)

Following World War II, the expanding economy of the United States was the engine of growth in the global economy. The United States set about restructuring the world economy so that the reemergence of Germany and Japan could drive capitalist growth and development in Europe and Asia.[1] The United States was virtually unchallenged in its global economic and political hegemony. The long boom (1950–1965) was followed by a period of fall in profitability and a turn from boom to crisis (1965–1973). A long downturn continued into the 1990s. The U.S. economy in the later periods was marked by slow growth in relation to accelerating expansion in Japan and Germany. The "intensified competition" between the three global blocs and "uneven development" led to overcapacity, overproduction, and a fall in profitability.[2] At the root of the post-war toxics crisis in the U.S. was the great economic expansion in competition with Japan and Germany and the Cold War military industrial complex.

The basis of the rapid growth of the U.S. economy following World War II was the chemical industry. By the 1960s, the United States was predominantly an urban nation, and a monopoly waste industry had emerged centered around two companies, Waste Management Incorporated and Browning Ferris Industries. Beginning their operations in Chicago, they colluded to take over a large portion of the waste industry nationwide. This industry, virtually unregulated until the 1980s, is an integral part of the U.S. economy, with the central sectors of the economy—banking, insurance, and big oil—all investing and sharing in the profits.

With the production and handling of waste such a major, profitable sector of the economy, the waste industry militated against needed changes, such as source reduction and recycling. This development is, in fact, exactly what one would expect in a capitalist economy in which policies are largely made and controlled by the largest and most powerful corporations. De facto "sustainable development" contradicts the inner logic of profits and accumulation under capitalism. The bottom line is profits and the necessity of achieving global competitiveness.

More than forty years of environmental regulation in the United States has resulted in a failure to protect the environment from toxic pollution and people's health from its effects.[3] The degradation of the environment should not be viewed as a failure to regulate effectively. Rather, the destruction of the environment is systemic: a necessary and integral part of the liberal-capitalist political economy of production, consumption, and waste. Seemingly ambitious attempts to protect the environment come down largely to window dressing, as large corporations control public policies in the protection of profits.

The extent of the threat of hazardous wastes to the environment and human well-being has become clearer over the past two decades as Congress has legislated to deal with the problem and attempted to devise strategies for cleanup and containment of hazardous wastes at the most contaminated sites. One indication of environmental crisis was the proliferation of Superfund sites—created and managed by the federal government to deal with the most hazardous toxic wastes—at the end of the 1980s. In 1987, the Government Accounting Office (GAO) reported to Congress that 70 percent of hazardous waste sites assessed by the Environmental Protection Agency (EPA) were leaking contaminants and that some 2,500 operating sites could require corrective action at a cost of some $22.7 billion.[4] Another government survey showed that "75 percent of permitted land disposal facilities were not in compliance with EPA requirements for groundwater, were leaking, or in a condition unknown to the agency."[5] Moreover, cleanup technologies were often merely cosmetic. Reports showed that 68 percent of remedies selected in 1987 failed to use any treatment whatsoever on the source of contamination. Forty-four percent of the remedies selected merely minimized exposure to contamination with fencing and capping.[6] Even after waste was removed from Superfund sites, 87 percent of landfills receiving Superfund waste were in an unacceptable condition. The EPA reported that most landfills would likely attain high failure rates shortly after fifty years of operation.[7]

Perhaps no area of public policy has created greater conflict between the interests of polluting industries, the public, and government regulatory agencies than that of the regulation of hazardous waste. The resulting "Superfund Syndrome" led to "a condition of constant confrontation among nearly everyone involved" and a failure to resolve one of the most critical conflicts in the United States.[8] Pollution continues, as indeed it must, while new methods of storage and disposal merely postpone the problem fifty to a hundred years down the road. Landfills remain toxic for hundreds or even thousands of years. Private corporations are increasingly given immunity from financial liability for their despoliation of the earth.

Such environmental destruction could not be kept a secret from the people forever. Due to an explosion in public awareness, greenwashing became a necessity for large corporations concerned about their public image. In the 1990s corporate interests in the United States launched a major public relations campaign to establish the perception that the crisis of environmental degradation from industrial pollution and other forms of waste had passed and that this was the "time for optimism."[9] Greenwashing is part of the new "corporate environmentalism," which touts "voluntary action" by corporations rather than government regulation and "command and control" confrontation. Such propaganda serves only to confuse and obfuscate the issues.

The crucial point, as John Bellamy Foster has pointed out, is that the history of capitalism has been a history of a war with the environment. Capitalism is a system of "creative destruction" driven by the logic of profit. This will not end except with a "socialization of nature and production" that will allow for "concern for other species and future generations."[10]

The Toxics Political Economy

The consumer revolution was seen as nothing less than the road to utopia. It depended on consumption and waste. The rapid expansion of the U.S. economy in the post–World War II period was largely based on the petrochemical industry as part of the emerging military industrial complex (MIC). Most of the more than 70,000 chemicals now used regularly have been produced since World War II.[11]

The emergence of the petrochemical revolution was marked by an upsurge in new miracle products, such as insecticides, plastics, synthetic rubber, rayon, dacron, and nylon—all derived from new products that were developed for the military in World War II. Petrochemicals, or organic feedstocks, were the bases for new products that began replacing "natural materials such as cotton, wood, rubber, metals, soap, manure, and natural solvents."[12]

The chemical industry, the "engine of growth" of the U.S. economy in the post war period, is its largest manufacturing sector, accounting for 10 percent of U.S. GDP from manufacturing and 2 percent of total U.S. GDP.[13] This industry has grown at roughly twice the rate of the U.S. economy since World War II and is characterized by very large industries. There are some 12,000 chemical manufacturing plants in the U.S. and 400,000 major chemical storage facilities.[14] The value of shipments of chemical and allied products in the U.S. rose from $268.1 billion in 1990 to $367.4 billion in 1996. The value of shipments of agricultural chemicals alone totaled $18.8 billion in 1992 and $23.4 billion in 1996.[15] Exports of chemicals were 17 percent of product shipments in 1995.[16] In the same year, U.S. shipments abroad were 13 percent of world exports.[17] Corporate profits were high in the chemical industry, averaging 13.2 percent for the five years prior to 1997 and 11.6 percent in 1997.[18]

The chemical manufacturing sector includes drugs and pharmaceuticals, which make up 25 percent of the industry, with $86.5 billion in products shipped in 1996. Industrial organic chemicals, the second largest component of the chemical industry, accounted for 20 percent of the industry, with $75.7 billion worth of products shipped. Finally, the plastics materials and synthetics components accounted for 15 percent of the industry, with $59.6 billion in products.[19] The U.S. is the world's largest consumer of chemicals, totaling $318 billion in 1995, not including plastics and resins.[20]

The chemical industry, as valuable as it is to the U.S. economy, comes at a high price. It is the largest industrial source of toxic chemical pollutants. It has played a "historical role" in the introduction of toxic substances into the production process.[21] As the source of toxics flowing through the economy, the chemical industry is the area where the issue of pollution prevention must be addressed.

The amount of toxic waste created by this industry is significant, as evidenced by the releases reported in the Toxic Release Inventory (TRI). For 1996, on- and off-site releases totaled 785 million pounds for 3,855 reporting facilities; production-related waste totaled some 10 billion pounds; non-production-related waste totaled 9.4 million pounds; other on-site waste management was reported as 8.3 billion pounds, while transfers off-site for further waste management came to 899 million pounds.[22]

Where did it all go? Fully half of all on- and off-site releases in chemical manufacturing were spewed into the air (392 million pounds); the earth was impregnated with a quarter of it (200 million pounds) in underground injection wells; 11.5 percent (90.4 million pounds) was dumped into surface water; other on-site land releases accounted for 8.7 percent, while 3.9 percent was transferred off-site to disposal. Some 2.4 million pounds were entombed in hazardous waste landfills on-site.[23] The largest transfer type off-site in the chemical industry was to "energy recovery," which means burning as fuel in cement kilns, boilers, asphalt plants, and so on. This totaled 378 million pounds. A reported 109.5 million pounds was sent to POTWs (Public Operated Treatment Works).[24]

What constitutes "hazardous waste" is quite complex. Definitions and criteria are set forth by the Resource Conservation and Recovery Act of 1976 (RCRA), the Superfund Law (CERCLA), and the EPA. But while objective, scientific criteria exist for classification, whether a particular substance is actually classified as hazardous is sometimes a political decision, depending on the added costs of production and handling of materials classified as hazardous to industries compelled to compete in the global marketplace.

Under RCRA, hazardous waste is defined as:

> a solid waste or combination of solid waste which because of its quantity, concentration, physical, chemical, or infectious attributes may (a) cause or significantly contribute to an increase in mortality or an increase in serious irreversible or incapacitating reversible illness, or (b) pose a substantial present or potential hazard to human health or the environment when improperly treated, stored, transported, or disposed of, or otherwise managed.

General RCRA criteria governing hazardous waste include toxicity, persistence, degradability, potential for accumulation in tissue, flammability, and corrosiveness.[25] It should be noted here that some wastes are considered hazardous but not under RCRA. These are special wastes, such as PCBs, asbestos, and radionuclides. PCBs and asbestos are regulated under the Toxic Substances Control Act (TSCA) and radionuclides under the Atomic Energy Act (AEA).

The definition of a hazardous waste under the Comprehensive Environmental Response, Compensation and Liability Act, or CERCLA, is broader. Specific contaminants include:

> any element, substance, compound or mixture, including disease causing agents, which after release into the environment and upon exposure, ingestion, inhalation, or assimilation into any organism, either directly from the environment or indirectly by ingestion through the food chain, will or may reasonably be anticipated to cause death, disease, behavioral abnormalities, cancer, genetic mutation, physiological malfunction (including malfunctions in reproduction), or physical deformations in such organisms as their offspring.[26]

According to the EPA, there are two ways a waste material is defined as hazardous: its presence on EPA developed lists of hazardous wastes ("listed wastes"); and evidence that the waste exhibits ignitable, corrosive, reactive, or toxic characteristics ("characteristic wastes"). Hazardous wastes include many substances from industrial processes, such as halogenated solvents, nonhalogenated solvents, electroplating baths, waste water treatment sludges, heavy ends, light ends, bottom tars, side cuts from distillation processes, and commercial chemical products when discarded. Among these chemicals are acute arsenic acid, cyanides, many pesticides, benzene, creosote, phenols, and toluene.[27] The EPA further divides hazardous wastes into five categories.[28]

Whether the EPA actually lists a substance as a "RCRA hazardous waste" depends on various factors—and here is where the matter becomes more complex, where definitions and criteria do not ultimately determine what is "hazardous." Any number of interpretations are possible; any number of exceptions permitted. A substance may be excluded by a "variance"—that is, be "delisted." Some commercial chemical products, for instance, "are not solid wastes if they are applied to the land in their ordinary manner of use or if they are fuels." Examples include pesticides when used as directed or fuel used in cement kilns, which would otherwise be classified as hazardous. Also excluded from classification as "hazardous waste" is household waste (although it is well established that it can indeed be hazardous); fertilizers from the growing of crops and animals; mining overburden returned to the mine site; fly ash, bottom ash, slag, flue gas emissions from the combustion of fossil fuels; wastes associated with the development of oil, natural gas, and geothermal energy; other wastes from leather tanning, waste water treatment sludges, and the production of titanium oxide; cement kiln dust; wood and wood products that failed the toxicity characteristics for arsenic; and petroleum-contaminated media and debris.[29]

Sometimes a hazardous waste may be classified nonhazardous by simply diluting it. "When the dilution rules apply, the mixture of a hazardous waste with the dilutent does not cause the dilutent to become hazardous and may render the hazardous waste nonhazardous."[30] In some cases dilution of hazardous wastes is permitted "for characteristic wastes and in those cases where the listed waste is not to be land-disposed."[31] And in some cases, hazardous wastes can be diluted for underground injection.[32]

While the EPA sets forth scientific and technological criteria for the classification of wastes as "hazardous," at some point what is treated as hazardous may well rest on political and economic grounds, rather than objective, scientific ones. A classic example is the mechanism whereby companies can petition to have wastes delisted, although these wastes are de facto hazardous. Whether or not a waste is classified as "hazardous" can be momentous in terms of the bottom line. Handling hazardous waste means significant operating expense for companies, while at the same time regulations open up opportunities for the waste companies doing the handling to make profits. For instance, Amoco planned to build and own its own disposal facilities while providing hazardous waste services to other companies. If delisted, de facto hazardous waste can be handled rather cheaply. It can be put in a non-hazardous waste landfill, or it can be used as fuel in a cement kiln.

Pivotal here—in this matter of delisting—is the "Hazardous Waste Identification Rule" (HWIR). The HWIR "identifies methods by which certain listed hazardous wastes, waste mixtures, and derived-from residue are no longer designated hazardous, either because they are generated with constituent concentrations that pose low risks or they are treated in a manner that reduces the concentrations to low levels of risk."[33] In one case, in December 1995, the EPA proposed that certain low-risk categories of hazardous waste not be included under subtitle C of RCRA. What is hazardous waste is not exactly fixed, not exactly pinned down. This is the point.

Those working in the "environmental industry" can often be straightforward about what is behind regulatory decisions at the EPA. As John F. Chadbourne, of Chadbourne Environment and Safety Programs, states, "Decisions on changes at EPA come to recognize the value of economic incentives to shape social policy, irrespective of incidental positive or negative effects on health and the environment." Health and the environment are not the top priorities of EPA policies, as Chadbourne reminds us. Social policy is driven by economic incentives. "The key factor is an acceptable economic return, which is essential to create the necessary incentives to solve material handling and operating problems."[34]

To comprehend what is happening to the environment, one must distance oneself from the notion that the EPA is about "environmental protection." Like most other bureaucracies of the federal government, it ensures that the interests of the U.S. corporate sector will be well served. If profits are large enough—for any number of related industries (waste producers, kilns using cheap waste as fuel, etc.)—then cement kilns will be de facto hazardous waste incinerators, spreading dioxins and other toxics in communities all across America. The waste industry, dealing with the excrement of industry as a lucrative commodity in itself, is at the center of capitalism.

Delisting, in general, clearly serves the interests of U.S. corporate capitalism. It gives corporations greater freedom to pollute without penalty, whether they send their de facto hazardous waste to nonhazardous waste landfills, dump it directly into the environment, or sell it for fuel. Even though hazardous waste companies might have made a profit on the handling of the waste if it had not been delisted, nevertheless, delisting serves the interests of most industries as it cuts production costs and allows them to compete more easily under globalized production. As a first approximation, delisting, giving firms greater freedom to pollute, tends to enhance the bottom line.

If what constitutes "hazardous waste" is ultimately determined by the interests of large corporations, whose political and economic power in the United States continues to grow by leaps and bounds, then it is equally true that the method of reporting toxic releases, the Toxics Release Inventory, is part of this same equation. Firms have often saved money by cutting corners, such as not reporting, misreporting, dumping hazardous materials illegally, or using hazardous materials illegally, such as in road building materials. The TRI was established under the Emergency Planning and Community Right-to-Know Act of 1986 (EPCRA). After passage of the Pollution Prevention Act of 1990, the TRI was expanded to include mandatory reporting of additional waste management and pollution prevention activities.[35]

According to the EPA, a "release" means "a discharge of a toxic chemical to the environment." Included are emissions to the air, discharges to bodies of water, releases to land (hazardous waste landfills), and to underground injection wells. Land releases may also include application farming (toxic sludge increasingly being used as fertilizer), surface impoundments, waste piles, spills, leaks, and RCRA Subtitle C (hazardous waste) landfills in future TRI reports.[36]

Data on toxic releases is collected according to the Biennial Reporting System (BRS). Whether reporting is required depends on the size of the company generating the waste. To report, the company must be a large-quantity generator (LQG); or must treat, store, or dispose of RCRA hazardous waste on-site in units subject to RCRA permitting requirements. A LQG is any site that generates 1,000 kilograms or more of RCRA hazardous waste in a single month; generates one kilogram or more of RCRA acute hazardous waste in a single month or at any time; or generates or accumulates at any time more than 100 kilograms of spill cleanup material contaminated with RCRA hazardous waste.[37] The situation is more complicated, however, as some states require that small-quantity generators report, and this data is included in the data for LQGs. Also the receivers of SQG hazardous wastes have to report their wastes, and so SQG waste should be included in the data reported in the National Biennial Report. This report includes only RCRA regulated hazardous wastes.

There are many limitations to the TRI. Through 1997, the program applies to industries in the manufacturing sector and those owned by the federal government only. Thus many sources of releases are not covered. Second, the TRI does not cover all toxic chemicals or all industry sectors. Facilities with fewer than ten full-time employees are not required to report. Also chemicals used in products such as pesticides and fertilizes are not reported or at most a small fraction of their actual release into the environment. The TRI does not account for toxic emissions from cars and trucks. Furthermore, facilities report "estimated data," and the program does not mandate that they monitor their releases. Different facilities use different methodologies for estimation. The EPA stresses that reporting requirements are being expanded and that many chemicals that are carcinogens have been added to the list. New industry sectors are being added to the list.

These developments might raise the question as to why the program did not use a more complete list of chemicals and require more sectors of industry to report earlier. The first reports required from RCRA Subtitle C treatment and disposal facilities (hazardous waste incinerators, landfills, and so on) came due on July 1, 1999, for the 1998 reporting year.[38] The terms used for reporting how the waste has been handled can be misleading unless one goes directly to the EPA publications to read the definitions used. For example, "used for energy recovery on-site" means that this quantity of the toxic chemical was burned as fuel in a cement kiln, boiler, or other such facility. The term "treated on-site" means the quantity of the toxic chemical destroyed in on-site operations such as burning in a hazardous waste incinerator.[39]

Given the various exclusions relating to what is defined as hazardous waste, the inconsistent and misleading methods of reporting toxic releases, and possible political maneuvering, numbers given for the disposal of hazardous wastes are not very meaningful, and it is quite difficult to interpret what the numbers really mean in terms of environmental degradation of the earth. Nonetheless, one thing remains clear: the amount of hazardous waste produced each day in the U.S. is alarming.

Clearly a toxic nightmare prevails in hundreds of communities who face corporate assaults of hazardous toxics in incredible amounts. These come through hazardous as well as solid waste landfills, hazardous waste incinerators, cement kilns burning hazardous waste, asphalt plants, and industrial boilers using toxic fuels. The costs of America's toxic nightmare are heavy. Air and water quality is unhealthy for at least 150 million. Drinking wells and ground water aquifers have frequently become contaminated with industrial chemicals, organic solvents, and metals.[40] No data is available on the toxic effects of 80 percent of all chemicals, and complete data exists for only some 2 percent or less. Occupational exposure limits have been set for only about 700 chemicals. Moreover, the EPA does not require testing for health effects before most chemicals go on the market.[41]

More than 200 industrial chemicals and pesticides have been found in the body tissues of 95 percent of Americans tested. The milk of 99 percent of mothers in the U.S. contains significant levels of DDT, and the average breast-fed infant ingests 9 times the permissible level of dieldrin.[42] And dioxin is found in milk, fish, meat, and paper in the United States as well.[43] Many chemicals, including these, now banned in the U.S., enter the country from Third World countries in the form of pesticide residues in imported foods. According to U.S. government reports, 25 percent of pesticides sold overseas by U.S. companies have been banned in the U.S.[44]

Government Regulation: Environmental Legislation

As Peter Montague has pointed out, no environmental legislation has ever been put in place that was not previously approved by the largest American corporations. Given the bottom line of profit making and capitalist accumulation, and the necessity of global competition, it is a foregone conclusion that environmental degradation is going to continue.[45] It is not a question of why the system of regulation did not work to protect the environment. The system did work, as it was supposed to, but its priority was to give U.S. industries the ability to compete with emerging nations abroad, where environmental regulation was often extremely lax or nonexistent. This was absolutely necessary if U.S. industries were to compete successfully with the emerging economies of Europe and east Asia and "win" the Cold War.

Beginning in the 1960s, Congress passed a series of laws nominally under the rubric of "protecting the environment." This legislation was prompted by a number of environmental disasters and new understanding accomplished by studies such as Rachel Carson's *Silent Spring*. The National Environmental Policy Act (NEPA) of 1969 was the first of 20 major federal environmental laws.[46] NEPA, for the first time, mandated that ecological considerations based on science must be a criterion for development, apart from economic and political factors. It was the basis for new legislation signaling that Congress intended to have a clear environmental bias, one that would be protected from capture by those business interests to be regulated.[47] NEPA required environmental impact statements and created the Council on Environmental Quality.

The new generation of legislation built on earlier laws. In 1955, Congress passed the Air Pollution Control Act, the first such federal legislation, which provided technical and financial assistance to the states. This legislation was prompted by a series of killer fogs, such as the London fog of December 1952, that had caused 4,000 deaths. It was designed to address catastrophic air pollution around areas of heavy smokestack industries.[48] Another London fog in 1962, killing some 700, helped prompt the Federal Clean Air Act of 1963. An air pollution crisis in New York in 1966 led to passage of the Federal Air Quality Act of 1967. The Federal Motor Vehicle Pollution Control Act of 1965, aimed at controlling vehicular emissions, was passed after a series of smog incidents in California. These efforts attempted to help states cope with the problem of air pollution but made no real progress.[49] The Clean Air Act of 1970 authorized greater federal involvement in air pollution control, but the provisions to regulate toxic air pollutants foundered on "scientific uncertainty." Under the complex system that emerged, regulators in the states defined regional emissions limitations in order to meet federal standards. Except for lead emissions, which declined by 65 percent in the first decade, total emissions of sulfur oxides, nitrogen oxides, and other particulates increased with the growth in motor vehicle use.[50]

Another area where legislation was overdue was in the area of safe drinking water. The Federal Water Pollution Control Act of 1948 subsidized publicly owned sewers. Heavy industrial pollution led to the notorious fire on the Cuyahoga River in 1969. Following this, the Federal Water Pollution Control Act Amendments of 1972 (Clean Water Act) contained two key provisions: first, all pollutant discharges to navigable waters must be eliminated by 1985, including specific toxic discharges; second, these waters must be fishable and swimmable by 1983. [51] An elaborate control system evolved, based on end-of-the-pipe control technology, under the National Pollutant Discharge Elimination System (NPDES).[52] This legislation was followed by the Safe Drinking Water Act of 1974.

In the area of toxic waste, legislation included the Toxic Substances Control Act of 1976, which authorized pre-market testing of chemical substances; allowed EPA to ban or regulate the manufacture, sale, or use of any chemical presenting an "unreasonable risk of injury to health or environment"; and prohibited most uses of PCBs. In 1976, Congress passed the Resource Conservation and Recovery Act, which required the EPA to set regulations for hazardous waste treatment, storage, transportation, and disposal and provided assistance for state hazardous waste programs under federal guidelines. In 1980, CERCLA, or Superfund, authorized the federal government to respond to hazardous waste emergencies and to "clean up" chemical dump sites; it created a $1.6 billion "Superfund" and established liability for cleanup costs. The Superfund Amendments and Reauthorization Act of 1986 (SARA) provided $8.5 billion through 1991 to clean up the nation's most dangerous abandoned chemical dumps; set strict standards and timetables for cleaning up such sites; and required that industry provide local communities with information on hazardous chemicals used or emitted. [53] The Pollution Prevention Act of 1990 attempted to make a major shift away from the end-of-the-pipe focus. By the 1990s, almost every state had established some form of waste minimization legislation, but a waste prevention framework still did not exist. [54]

The EPA: The "Super-Regulatory" Agency

By the end of the 1960s, the emerging crisis and increasing environmental awareness among the public led to the push for a super-regulatory agency that would coordinate efforts toward "environmental protection." The complex command and control governmental machinery put in place in the coming decade, however, was certain to run up against the demands and inherent logic of the capitalist system itself, the need to compete under an emerging system of globalized production as well as the necessity of servicing the needs of investors for dividends and profits if the industry were to remain viable.

With the coming of the Reagan administration in the 1980s, the corporate revolt against regulatory threats—namely, anything that limited the freedom of industries to cut production costs in order to maintain a competitive edge in the marketplace and to increase profits and accumulation—resulted, predictably, in the capture of governmental regulatory machinery by big corporations. Giant waste-producing corporations were now essentially in control of the agency that was supposed to regulate them; wide-reaching revolt against governmental regulation during the Reagan years meant that further attempts at command and control were slated to fail. With the predictable triumph of corporate money and lobbying in Washington, an era of "corporate voluntarism" was to emerge in the 1990s. The corporations would "regulate" themselves.

In 1970, President Nixon created the Environmental Protection Agency as a separate federal agency. [55] The agency was launched in January 1971 with a staff of 7,000 and a $3 billion budget under director William Ruckelshaus. Twenty years later, in 1991, the agency had 17,000 employees and a $6 billion budget with departments to deal with air, noise, radiation, water, solid waste and emergency response, pesticides, and toxic substances. [56]

The EPA was designed to avoid political capture, but clear evidence of political control of the bureaucracy, not surprisingly, has been documented. [57] Executive appointments and budget powers reduced both EPA abatement actions and civil-case referrals under the Clean Water Act. Sometimes EPA bureaucrats were able to use more subtle strategies and "hidden actions" to maintain a strong enforcement action against Reagan administration wishes. The Reagan administration was able to use the White House and the Office of Management and Budget (OMB) to largely delay EPA enforcement action under Superfund for the early years of the Reagan period. Anne Burford, chief EPA administrator, and Rita

Lavelle, in charge of Superfund, blocked cleanup action.[58] In December 1982, a House investigative subcommittee subpoenaed 700,000 EPA documents related to Superfund. This investigation led to Lavelle's later conviction of perjury for her testimony. Environmentalists were angry that the EPA was stalling on cleaning up toxic waste sites. A backlash against Reagan administration policy occurred after the "Sewergate Scandal" [59] and Burford resigned in 1983.

A number of analysts have concluded that the failure of U.S. environmental legislation stemmed from a classic case of "disjointed incrementalism" [60]—that is, planning without comprehensive goals, a lack of coherence in implementation, and a pollution control agenda that was not implementable. [61] Several dysfunctional problems plagued attempts by the federal government to regulate the environment. First, the focus on a single-medium approach—air, water, and land—led to merely shifting the waste from one medium to another. Second, the focus on pollution control rather than prevention led to end-of-the-pipe remedies and technologies that did not deal with the root of the problem: production of the waste itself. A pollution control industry evolved to treat wastes only after they were produced. Third, policy focused on treatment, storage, and land disposal of hazardous solid waste. Fourth, a focus on "technocentrism" became infused in the bureaucracy, institutions, and constituencies. A strong public belief in technology militated against any initiatives to move the policy-making process upstream to production processes and pollution prevention. [62] Fifth, in the 1980s, under Superfund, economic costs of pollution control became increasingly prohibitive as liability and litigation became institutionalized. All this led to merely shifting the location of pollutants. There is some truth in this analysis. But the overall tendency is best explained by the fact that corporate profits and global competitiveness required production techniques that produced and dumped more waste into the environment.

Preventing pollution has been seen as the "challenge of environmental policy making" in the future. [63] In 1989, EPA endorsed this approach with a statement in the *Federal Register*. Change would include, first, multimedium as a basis for regulation, and integrated pollution control, upstream; second, it would involve a new relationship or partnership between regulators and regulated, away from the "command-and-control" format.[64] Part of this agenda was a response to demands for greater market forces. This plan led increasingly to a system of trade in pollution rights and a range of policies based on corporate initiatives to monitor their own performance. Such practices are more easily accomplished by corporations operating on a global basis, as long as they are free to move waste abroad, or more likely to incinerate it on-site as "energy recovery."

Some analysts have argued that problems of public policy are deeply systemic in an industrial society and grounded in "technocentrism," the deep faith that technology ultimately works—that it is just a matter of finding the right technology to solve the problem. Engineers, it is argued, tackle the problem at the wrong end of the stream, dealing with pollution rather than with the source of pollution upstream in the production process. The problem is indeed systemic, but entailed in the logic of capitalist production itself. Technology under capitalism is designed not to protect the environment but to manufacture products cheaper—to beat out the competition. Source reduction of toxics can sometimes be achieved to an extent, but not at the expense of the bottom line of corporate profits.

By the 1970s, legislation did provide for the possibility of policy options that might help prevent pollution, but, again not surprisingly, these did not emerge. In the case of automobiles, for example, State Implementation Plans (SIPS) could have utilized such tools as parking controls and surcharges, auto-free zoning, mandatory bus and carpool lanes, bridge and highway tolls, vehicle driving bans, four-day work weeks, gas surcharges, and rationing schemes.[65]

The Toxics Substances Control Act has been called the "story of the road not taken" with the EPA restricting only 5 chemicals in the first 15 years (PCBs, CFCs in aerosols, nitrides in metal working fluids, dioxin and asbestos).[66] The road was "not taken," clearly, because it threatened corporate profits. The EPA ran up against the Confidential Business Information (CBI) provisions of the act, which allowed manufacturers to withhold information on the basis of trade secrets. This CBI claim was made on 90 percent of new chemicals, resulting in virtually useless data, while a 1992 study found blatant abuses by companies. This attempt to regulate toxic substances foundered on the assumption that the "bad chemicals" must be found, when the real problem was the toxicity in the use of thousands of chemicals.[67] This oversimplified, cosmetic approach is clearly useful to U.S. corporate industry as it gives the impression that the federal government is doing its job to protect the public from dangerous toxics, while failing to get to the root of the problem. The toxic economy remains in place, and new toxic substances proliferate.

The effort to move toward hazardous waste incineration was grounded in congressional concern about land disposal hazards, a way to "minimize waste."[68] Yet waste incineration not only contributes massively to "land disposal hazards" but to hundreds of contaminants in the air and dioxin in the food chain as well. The Hazardous Solid Waste Amendments to RCRA (HSWA) in 1984 made the reduction of hazardous waste "a matter of national policy." Incineration was considered an "advanced treatment technique" and an alternative to land disposal of wastes, even though that is where the ash from waste incineration goes.[69] So the effort was on reduction through treatment rather than prevention, but treatment often comes down to burning hazardous waste in cement kilns and industrial boilers. It remained cheaper to dump solid waste in distant landfills and burn hazardous waste as fuel, while recycling part of it.

The Pollution Prevention Act (PPA) again aimed at reducing pollution at its source but was a compromise with industry. First, the law would retain the requirement for "source reduction" in exchange for more corporate voluntarism; second, reporting requirements of the TRI were tightened, in exchange for the abandonment of efforts to address CBI trade secrets abuses. Source reduction meant either closed-loop in-plant recycling or off-site recycling. Some 90 percent of source reduction was in-plant recycling (and sometimes not real recycling). More toxic wastes began to be burned on-site as fuels.[70]

The Pollution Prevention Office (PPO) of the EPA too was largely a token effort, with a staff of only 15 after two years. The 33/50 Program to reduce the release of 17 toxic chemicals appeared to be more serious, if only a drop in the chemical bucket. Industries would reduce TRI release chemicals by 33 percent by 1992 and by 50 percent by 1995. This was a voluntarist approach, and the 17 chemicals represented 22 percent of the TRI releases reported in 1988 (1.4 billion pounds). These were, not surprisingly, the 17 that seemed easiest to reduce, meaning cheapest to reduce. The program was clearly designed to protect corporate profits while giving the appearance of seriously dealing with a number of dangerous toxics. Methods to "reduce toxics" often did not represent a form of pollution prevention; reduction could be accomplished through condensation or distillation. A total of 812 companies participated, but the effort was often deceptive; there were "phantom" or paper reductions, and many touted achievements were, in fact, release reductions rather than source reductions. [71] It was easier to bring companies on board, however, with incentives. Big companies smelled new profits.

New Federalism

The policy of New Federalism (NF) emerged in the seventies and eighties. Under the principle of "partial preemption," the federal government sets minimum environmental standards and objectives, and the states may implement their own laws. [72] "Preemption" is a fancy term for stripping local people of whatever democratic tools remain at their disposal for fighting toxics in their communities. So-called New Federalism began with the Local Fiscal Assistance Act of 1972, with less federal funds given to states. The Reagan administration approach was to decentralize and defund federal environmental protection activities, including those dealing with hazardous waste. Many states were not able to assume environmental responsibilities. Some did have innovative programs. Massachusetts, for instance, worked to remove pollution at the source. Florida operated a hazardous waste collection program, while Wisconsin designed a stationary source air pollution control program, conducted extensive research on acid rain, and developed a toxics program. Regulatory efforts primarily relied upon permits for polluters—the rules being proposed by the Department of Natural Resources (DNR), aired in public hearings, and subject to possible veto by the state legislature. The program was said to "enable rapid responsiveness" by the state to air pollution problems. Delaware's management of extremely hazardous substances was another "innovative program."[73] States cannot effectively manage toxic wastes, however, without enormous sums of money and highly trained staff. But under CERCLA, states were required to pay 10 percent of cleanup costs or 50 percent if the site was owned by the state or a municipality.[74] Such "federalism" would obviously serve to benefit corporate profits in most states by forcing the taxpayer to pay corporate costs. Federalism in the control of waste was strictly prohibited under court rulings dealing with the Commerce Clause of the U.S. Constitution. Allowing private waste companies to move waste on an interstate basis contributed to "economies of scale" and a wave of mergers in the waste industry, accelerating the trend toward monopoly. The largest industries combined into nationwide or regional operations, while the smaller companies were increasingly gobbled up by the bigger operators. At the same time, private corporate control by the emerging giants stripped local communities of the control of their own waste.

From Command and Control to Corporate Voluntarism

The command-and-control approach to regulation pursued by federal government agencies involved four steps: Rules and regulations on control of pollution sources must be determined; a set of penalties or sanctions to be imposed for noncompliance must be established; sources must be monitored for noncompliance or a system of self-reporting established; violators must be punished usually through administrative or legal means.[75]

The failure of the command-and-control model of regulation led to a number of economics-based initiatives during the Reagan and Bush administrations. Among these are tradable discharge permits (TDPs), bubble transactions, offset transactions, and emissions marketing and banking.[76] Some economists have argued that the major disadvantage of command-and-control regulation is that it is too costly and provides only negative incentives. In this view, there is a weak incentive for polluters to comply with regulation since it is highly uncertain what the consequence of noncompliance will be.[77] Market-based initiatives were strengthened by President Reagan in February 1981 by Executive Order 12291, which mandated that major federal regulations be subject to an economic analysis of costs and benefits.[78]

Many economists argue that market-based initiatives ensure the greatest benefit for cost, resulting in the most efficient use of resources in society. This proposition, often taken as an article of faith, is, in fact, an empirical question. Cost-benefit analysis sometimes results in policies that tend to benefit the environment, but not necessarily, as the negative externalities of environmental degradation are often ignored. On the positive side, cost-benefit analysis resulted, for example, in the regulation that maximum lead content in gasoline be reduced from 1.1 grams per gallon to 0.1 gram. In another instance, cost-benefit analysis led to the decision not to build dams after the Army Corps of Engineers had exaggerated the benefit of the facilities.[79]

Applied to environmental policy decision making, however, cost-benefit analysis cannot, in general, be an adequate tool for the protection of the environment. This is because some monetary value must be put upon living nature and the ecology. That is, nature must be treated as a commodity. In the first place this is a highly subjective exercise because no one knows what the monetary value of nature really is. Within the liberal capitalist framework, when the system must generate profits and compete with other economic blocs within the framework of nationalistic competition, it is quite predictable that environmental degradation will continue. In fact the preponderance of evidence bears this out beyond any doubt.[80] Environmental trends, based on 21 indicators of environmental quality, provide strong empirical evidence that 20 years of environmental policy making has been taking the earth in precisely the wrong direction. And at what cost?

In 1979 the EPA wrote regulations allowing industries to buy, sell, and bank, for later use, tradable discharge permits. [81] Congress has also authorized plant owners to purchase additional pollution allowances from the EPA. These have been applied to the reduction of landfill gas emissions. Under the bubble concept, an entire manufacturing facility, rather than a single polluting stack, is considered a point source. This arrangement allows plant managers to manage activities to control the aggregate discharge of the facility and lower costs of meeting emissions requirements.[82] By 1990, the EPA had approved over one hundred bubble transactions.

Offsets are trade offs that allow the establishment of a new source of pollution in a non-attainment area—one that has not met federal air quality standards—by getting another source in the same area to reduce its pollution load by a greater amount. For example, in an industrial smokestack area, where the air is already dirty, offsets allow a company to close down an old factory and open up a new one in the same area as long as the new factory does not produce more emissions than the old one. Offsets also allow plants to save money by increasing emissions from those stacks of a facility that would be expensive to clean up while using equipment to reduce emissions from stacks that are relatively inexpensive to clean up, as long as the overall volume of emissions remains the same. This simply means that while the volume of emissions may remain constant, the toxicity of the emissions may increase greatly. [83] Consequently, bubble transactions and offsets allow companies to meet the requirements of the law and save money while sometimes actually putting more poison into the air.

Such laws, then, would seem to be written in the interests of the companies they are supposedly intended to regulate. By 1990, the EPA had approved over 2,000 offsets, about 10 percent between firms.[84] Offsets, the argument goes, encourage older plants to shut down. Even if offsets tend to encourage companies to close old plants more quickly, new plants are opened to replace them, which guarantees no overall improvement in air quality. The legislation simply legalizes and institutionalizes the condition of dirty air ("nonattainment areas" to the government technocrats). Moreover, the pollution of relatively clean areas—that is, "underpolluted areas"—is encouraged under the Prevention of Significant Deterioration (PSD) concept. The policy proposes that these "underpolluted areas" be allowed gradually to increase their pollution levels. Some states have also been able to get around the federal Ambient Air Quality Standards by building very tall stacks so that the emissions are dispersed so high in the air that they are never measured. This is because readings of the federal Ambient Air Quality Standards are measured close to the ground. The states simply write into their State Implementation Plans (SIPs) that tall stacks will be used. These tall stacks then disperse the pollution high into the clouds, but also increase acid rain, as the pollution is sent on to other states.[85] The trend among policy makers seems to be that the EPA should be left free to bargain with and get what it can from industries.

The failure of the command-and-control approach, quite predictable in the capitalist system, is most clearly demonstrated in the case of Superfund. The Superfund dilemma was essentially a conflict over who would pay the $100 billion toxic debt and was at the root of the political struggle over the potential threat that it posed to corporate profit. Under capitalism, toxic liability must accrue to the public sector, not the private. In America, the largest industries have little to fear, except from a potential grassroots swell of opposition to the dumping of toxics in local communities. "Is it possible to successfully pursue environmental goals in a political economic system that is biased toward the interests of capital?"[86] is the question one scholar poses. The interests of capital are arguably more than a "bias." Profits are "the law and the prophets" of the capitalist system. Certainly environmental goals can be pursued. They are being pursued as a form of class struggle against the most powerful forces in America by ordinary people in local communities. Superfund is a crucial illustration of market failure in a system where the major corporate polluters have enormous economic and political power.[87]

By 1980 studies of hazardous waste sites left no doubt that the crisis was very great. Studies documented that there were some 30,000 to 50,000 hazardous waste sites; at some 12,000 to 34,000 of these, there was significant surface and groundwater contamination. The EPA assessed 8,000 industrial sites and found that 70 percent were unlined; 95 percent were virtually unmonitored for groundwater contamination; at least 50 percent contained constituents that would contaminate groundwater; 30 percent were located above usable aquifers that would likely be contaminated by leachate from wastes; and 33 percent were located within one mile of water supply wells that would be contaminated by groundwater flowing from the site.[88] When groundwater is contaminated by toxic chemicals, it remains toxic for hundreds or even thousands of years.[89] Barnett has estimated that at least 9.24 million people were at risk from contamination of surface water and 24.2 million from groundwater.[90]

Superfund (CERCLA) attempted to make polluters the responsible parties, and liable for the cleanup, although it was estimated that not more than half the cost could be recovered through enforcement. "Joint and several liability held any responsible party potentially liable for the whole cost of cleanup even if others were also responsible."[91] But CERCLA (PL 96-510), signed into law on December 11, 1980, by President Jimmy Carter, came into effect just as the antiregulatory atmosphere of the early Reagan administration was taking hold.

The EPA was to put 400 sites on the National Priorities List. Early on, however, President Reagan ordered that OMB had to clear all regulatory actions. Rita Lavelle, who headed Superfund, came to the administration from Aerojet-General, which had the third worst pollution record in California. Robert Perry, EPA General Counsel, was a former lawyer for Exxon.[92] Reagan administration cuts in the EPA, from 1981 to 1983, included a 33 percent cut in total budget and a 39 percent cut in the hazardous waste program, in real dollars; a 21 percent cut in EPA abatement and control staff; a 33 percent cut in enforcement staff; and a 16 percent cut in research and development (R&D) funds.[93] The Sewergate Scandal resulted in the firing of EPA director Anne Burford in March 1983 and her replacement by William Ruckelshaus. Under the new director, the cleanup program was put on track, but the "transfer costs," in litigation alone, were prodigious. During the first eight years of the program, $4 billion went to contractors for services and 40 percent to lawyers, and the program came to be known as a "welfare program for lawyers." [94]

In October 1986, the Superfund Amendments and Reauthorization Act was passed. Attempts were made to address the issue of compensation of those with injury and disease from hazardous wastes, but any such provision was strongly opposed by insurance companies who feared potentially enormous liability. Instead, insurance companies argued that victims should sue under tort law in state courts. Securing compensation in court for the victims, however, would be a very long shot, indeed. Why? Because even though the aggregate evidence of the association between hazardous waste and health problems is overwhelming, such evidence is generally not accepted, in courts, as proof of causation.[95] To qualify for Superfund-financed cleanups under SARA, states were required to have sufficient capacity for the disposal of all hazardous waste expected to be generated within the state for the next 20 years. [96]

In 1987, the Bush administration OMB lowered health and cleanup standards. While the EPA had used a limit of one cancer per 10 million exposed to a contaminant, the OMB set the limit at one cancer per one million exposed. [97] Reduced standards for cleanup included lowering the upper-end risk range for carcinogens; considering hazardous waste facilities "cleaned up" when contamination problems beyond the facility borders were addressed, the hazardous waste within the borders remaining; scrapping of stringent remedy selection standards; treating several contaminated units within a facility as a single unit; and allowing waste to be transferred between units without being subject to the RCRA ban on land disposal of untreated waste.[98]

In the late 1980s, the trend toward "socializing" the toxic debt accelerated; ultimately the taxpayer would pay, either now or later. Large corporations, such as Du Pont, Rohm and Hass, and Texaco, began to sue municipalities for a share in toxic site cleanups under the rationale that the cost of capping sites is related to the volume of waste deposited, regardless of toxicity.[99] They argued that the cities should share the cost of capping a toxic waste site, because even if cities had not dumped toxic waste or had dumped waste that was less toxic than that of the companies, they nevertheless increased significantly the volume of waste in the dump and consequently the cost of capping the site. SARA made further inroads in dumping the toxic debt in the public sphere when it opened the door to laundering schemes to cleanse companies of liabilities and saddle taxpayers with the cleanup bill. This policy goes forward today under "Brownfields legislation." For example, SARA provided for an innocent landowner defense whereby, in case the purchaser was unaware the site was contaminated or in the case of foreclosure, the present owner would not be held responsible. This solved the problem of selling old industrial sites for reuse when the potential hazardous substance liabilities on some properties greatly exceeded the market value of the property thus presenting impediments to financial and real estate transactions. Brownfields legislation resulted from pressure from businesses for new laws that would provide exemption from liability, "a major loophole allowing liability to be extinguished in the transfer of property through a bank. Laundering schemes, it was feared by some, could cleanse properties of liability and leave Superfund with the cleanup bill."[100]

The enormous costs of "cleanup" have meant that not only is the private sector resisting spending for this purpose, but increasingly the public sector as well. Both government attempts to regulate and the Superfund cleanup have failed, but through greenwashing, if the battle cannot be won, all concerned can declare a victory and pull out.

Corporate Greenwash

Greenwashing strategies present an essentially upbeat message, a positive picture of the environment and the political and economic policies governing it. Greenwashing charges "radical" environmentalists with pessimistic thinking, with distortions and outright misrepresentations with respect to the issue of environmental degradation.

One argument of mainstream economists is the "growth-first" hypothesis, which claims that economic growth is a prerequisite for environmental protection and that environmental protection improves with economic growth. The other side of their argument is sustainable development. Both arguments need to be examined.[101]

According to the development-first thesis, most environmental dangers are in the Third World. These include dung smoke in cooking, dirty water, and so on. The effects of PCBs, radon, Alar, and pesticides pale beside the environmental problems that accompany underdevelopment. So, according to this logic, the Third World must be saved by Western technology.[102]

This tack ignores the actual impact of Western technology in much of the Third World, for example the production and use of pesticides, asbestos, dams, petroleum refining, mining, etc. The worst manufacturing processes, once the harmful effects are found out, are often curbed and banned in the West but quickly smuggled off to the Third World to continue their environmental degradation. As soon as McDonald's agrees to stop selling its McToxics Styrofoam in the U.S., it moves to selling the same packaging to the relatively unaware all across urban areas of Turkey and other emerging markets. As they are curbed in the U.S., PVC plants spring up in developing countries to replace any environmental degradation that might have been prevented by activism in the U.S. This globalization of corporate toxic pollution has yet to be adequately addressed and signals the need for global campaigns against toxics. Still the barriers are many, particularly in poorer countries where urban migrants are just new entrants to the consumer revolution.

A key concept that emerged from the Brundtland Commission Report, the document produced by the U.N. World Commission on Environment and Development, is "sustainable development." This concept replaces the principle of maximizing utility in the conventional development framework. Sustainable development is based on two propositions: First, economic and social development should enhance rather than degrade the resource base. Second, interventions designed to achieve development must not compromise the ability of future generations to meet their needs and aspirations.[103] In reality, however, "sustainable development" must be read as "sustainable profits" since the latter is generally the desired thrust of those who employ the term.

There are several fundamental difficulties with the concept of sustainable development. First, poverty is seen as largely responsible for environmental degradation, and therefore economic growth is needed to alleviate poverty. In focusing on the "need for development," rather than inequitable global distribution, the concept, as well as the report, follows the growth-first hypothesis.[104] If growth were to be redefined to "mean an improvement in human welfare, rather than increasing production for consumption,"[105] the focus could be shifted to the underlying need for global political and institutional structural reform.[106] Some charge that SD is only a "comforting incantation," also suggested in the phrase "sustainable increase in the rate of economic growth."[107] In fact, "sustainable development" is more than a comforting incantation. It is an element in the ideology of liberal capitalism that tends to conceal the continued environmental degradation and the underlying neglect of the environment that is necessitated by the inherent logic of the capitalist system itself and competition under globalized production. What is also concealed is the question of whether "sustainable development" is to be grassroots democratic development that serves local communities or growth that serves to increase the profits of major national and global corporations while depleting local communities of environmentally friendly capital. There is nothing wrong with the concept, as such. The real problem is that it is not possible to take it seriously under the current logic and operation of global corporations.

Another greenwashing approach is to argue that the public has been badly misled by environmentalists, who are not really concerned with protecting the environment but rather with making a career for themselves as environmentalists and lobbyists. Greens are charged with being purists and elitists, who cannot tolerate the good news that the worst of the environmental crisis is over.[108] This is the "new environmentalism." To promote it, corporations spend some one billion dollars a year to make themselves appear to be "green" and at the same time fight a war against environmental activists. There are now dozens of green-sounding, but spurious, pro-corporate environmental NGOs. The movement can be traced to 1934, when Du Pont formed a public relations department after a Senate investigation of the gunpowder industry charged Du Pont with being a merchant of death. Metropolitan Edison, after Three Mile Island, and Exxon, after the *Exxon Valdez* oil spill, followed this same strategy.[109] After the Times Beach, Missouri, incident, there was a campaign to "detoxify" dioxin.[110] Dr. Vernon Houk of the Centers for Disease Control said later that his evacuation of the town in 1982 had been a mistake.

Under such greenwashing strategies, 3M Company launched its 3P program, "Pollution Prevention Pays"; Chem Waste Management was renamed WMX; and Technologies and Nuclear Engineering, Inc., became "U.S. Ecology." Amoco created "Ecova" Corporation to build the hazardous waste incinerator in Kimball, Nebraska, now owned and operated by "Clean Harbors of Braintree, Inc." Dow Chemical, 3M, and Polaroid are said to be "well-run companies." 3M's claim to have prevented 72 million pounds of pollution every year from 1975 to 1989 is misleading. Because of the increase in production, total output of pollutants actually increased in that period.[111] These companies remain among the nation's worst polluters. Under another program, billed the "Green Lights Program," oil, utilities, and petrochemical companies install energy-efficient fluorescent bulbs.[112] There are also many active corporate NGOs, such as the American Nuclear Society and the Asbestos Institute of Canada. These companies portray a "win-win fantasy" to the public.

Another greenwashing strategy is to play on "uncertainty." For example, greenwashing accepts ecological degradation but not an environmental crisis. The Bush administration strategy was to raise scientific uncertainty, through such organizations as the right-wing CATO Institute, about such issues as global warming.[113] As Lamont C. Hempel observes, "politics dominates when science equivocates."[114] This is what many corporations know and bank on.

This confusionist approach is seen in Greg Easterbrook's article "The Illusions of Scientific Certainty," published by Gale Research in 1993.[115]

Gale Research is associated with the Coalition for Environmentally Responsible Economics (CERES). The arguments in Easterbrook's article are those commonly used by big business in their attack on environmentalists. They are standard, all-purpose arguments.[116] While arguing for better "science," Easterbrook attacks scientists who have documented the deadly effects of PCBs and dioxin. "Today there is a body of science suggesting that dioxin, though a toxicant, is not the deadly substance scientists once thought," he writes.[117] The scientific reports that favor industry, not surprisingly, find great favor with Easterbrook. In attacking the way public policy and environmental regulations are made, he complains that policy is influenced by oil spills and unaware journalists. So scientists are left out of the loop and do not play the role they should. He does not stress the extent to which public policy is driven by corporate lobbyists, who have inordinately more political clout than environmentalists.

A crude attempt at confusionist rhetoric is evident in Michael Fumento's *Science Under Siege*, published by William Morrow. According to the book jacket blurb, "Fumento proves conclusively that dioxin, video-display terminals, power lines, pesticides, and other products are not the deadly threats to you and your family's health that you have heard, while some touted solutions to the real environmental problems, such as fuel gasohol, are grounded not in good science but in cynical, dollar driven politics." This statement says much about the confusionist approach of the book. The final chapter, entitled "Ending the Reign of Terror," draws an analogy between the "environmental revolution" and "the Reign of Terror" in the French Revolution. The charges against environmental "terrorists" include the "persecution of chemicals, processes, devices, and manufacturers," "squandering" the scarce resource of money in the name of the environment, and bringing down the real GDP of the U.S. by 2.6 percent with the environmental regulations enacted before 1990.[118] The argument is made that regulations have rendered it difficult for the U.S. to compete with other nations. Americans must get beyond their "irrational fears" and understand that "[E]ating vegetables and fruits treated with pesticides made by Dow and Uniroyal is incomparably safer than downing the delicious ice cream produced by the environmentally and politically correct company Ben and Jerry's, chock-full of saturated fat . . . What is needed is to end our Reign of Terror, to restore sanity and sound principles to our revolution."[119]

Still another greenwashing strategy is to attack the press. Reporters and journalists come in for some of the hardest hits from Easterbrook, although he himself is a journalist. News organizations and politics "want panic." They cannot be cool, rational, and scientific. When a scientist explains a point, "the eyes of reporters glaze over."

Justin Dart, the late California business magnate and backer of Ronald Reagan for governor of California, stated that "environmentalism is the disease of modern America."[120] It is from this "disease" that the corporate attack on the environmental movement seeks to rescue America. The future looks rather bleak for those concerned about the environment, considering the number of large, expensively produced, authoritative-looking greenwashing "environmental publications" filled with industry arguments being produced with corporate money and provided to public libraries across the country. It is not difficult to see what environmentalists are up against with publishing going to a corporate-controlled Internet system. The Internet can be a powerful global corporate propaganda tool for saturating the world with corporate greenwashing propaganda and public relations industry lies. Today, those who would work to save local resources are up against both powerful global institutional structures and a powerful ideology, not to mention the technology to spread it to every nook and cranny of the globe. This is truly the "Age of Greenwashing."

2
Wasting America: Capitalism, Waste, and the Market in the United States

Private waste haulers are after this commodity called "waste." Everybody wants it because there's money involved.
— Ron Weatherman

Under monopoly capitalism the rapidly growing power of global corporations is stripping citizens of democratic power fought for and won historically. The fundamental driving force today is the growing power nationally and internationally of the contemporary corporation, essentially a totalitarian form of organization. As states and governments become weaker and hollowed out, corporations, accountable only to the bottom line, increase their power, increasingly stripping people of their democratic rights. Ironically, the stupendous global propaganda machine, led by the United States and global business media networks, is successful in projecting a vision of freedom emerging from privatization, globalization, and the retreat of the state. Nothing could be further from the truth. A point of departure for examining this trend is the growing power of large U.S. corporations in continuing the process of rolling back environmental regulation and continuing business as usual in the area of toxic pollution.[1]

To put this in a historical context, one can conceptualize the process as a deepening and continuation of the historical process of enclosure.[2] The enclosure movement, as Marx conceived it, laid the basis, largely through force, for the emergence of capitalism. This process worked to deprive a large portion of the population of the means of subsistence. In this way, a propertied class was engineered, through fiat, at the same time creating a class that lacked property, but more importantly lacked any power to live on available resources as a result of enclosure. So the class forced to work, driven into wage labor, was created as a necessary basis for the emergence of capitalism. It is a process that continues today in different forms, whether in the Midwest of the U.S. or in the Global South. It is at the root of historical environmental destruction.

In the United States, environmental "protection" is embedded in a political and economic framework that enables corporations to maximize profits and capitalist accumulation. Within that framework, corporations seek out "opportunities" to increase profits and invest accordingly. They do not oppose environmental regulations and legislation in every case. Waste companies even support stringent regulations when they believe these regulations will create a market for services they can provide for a profit. As we have emphasized, however, as a first approximation, deregulation and the maximum freedom for companies to operate without regard to environmental degradation is in the interests of U.S. capitalism. With their ability to provide the capital necessary to meet more stringent requirements, monopoly solid and hazardous waste corporations have profited from new subtitle D landfills and new hazardous waste regulations. Besides these regulations, a number of other mechanisms exist that contribute to capitalist accumulation and monopoly power in the waste industry. One powerful device is the Commerce Clause, which has been and continues to be used to force waste facilities onto local communities, stripping people of democratic control of their environment. Furthermore, corporations operate on a global level. The current thrust of "open markets" is designed to allow global corporations to produce and dump their toxics anywhere.

The Waste Industry: Wasting Products and Workers
The waste industry is at the very center of capitalist production, revealing in a particularly stark way the underlying logic of the system to produce, sell, and utilize as little as possible what use values reside in the commod-

ity while maximizing the exchange value. In exploring the reason for the enormous waste of infrastructural capacity, or the "built environment," Stephen Horton observes that capitalism has a structural preference for exchange value over use value. This is because it is through exchange that the surplus value contained in the commodity is valorized, or accrues to the capitalist. Horton has shown how capitalism comes to waste use value, such as in the rapid demolition of usable buildings to create post-modern inner cities.[3] The same observation may be made about solid waste: the less use value the consumer obtains from the products he or she buys, the better it is for capitalism as long as the consumer keeps buying more. The more dumped into landfills, the more profits realized.

Big waste companies own something else that is extremely valuable. Due to RCRA Title D, they own the only places in the Unites States where society can legally store its solid waste. The same is true for the ownership of RCRA Title C landfills for hazardous waste. When society pays monopoly prices for a waste company to take its waste, it is, essentially, paying for a storage space for that garbage. How long it will be stored there, no one knows. Since the waste industry has a monopoly on the only legal spaces for storage, and since it costs millions to build a legal Subtitle D landfill, society has little choice but to pay the going rate, and this is often monopoly prices.

So exchange in the waste industry is a reverse exchange. Waste companies make large profits in storing waste and set up the conditions for future Superfund sites, even while toxics pollute the groundwater. At the end of 25 years of operation and another period of "monitoring," they walk away from their landfill with their profits, and the toxic liability, the massive externalities, reverts to the public sector. The public now gets the waste back that it paid the waste company to store for some 20 to 50 years. The waste industry has already sold the marketable portions of the stored waste, depending on the materials markets, the needs for materials, public subsidies and so on. Further subsidies and credits are gained from the use of methane produced in landfills. In other words, consumers are buying back what they have paid the waste companies to take and store for them.

Since commodities are generally discarded before their use value is used up, from a Marxist perspective, waste is a store of use value. The use value that waste contains was created in the labor process, just as in any other commodity. It is "congealed labor." In other words, waste is created by the metamorphosis of materials and the labor of workers into the product. When waste in the form of unutilized use values is stored in a waste dump, what is stored is the blood, sweat, and tears of laborers, of the working class. In a very real sense, it is the workers themselves.

Under capitalism, as analyzed by Marx, since the working day is divided into "necessary labor" and "surplus labor," the worker must work longer than is necessary to produce his living. The additional time produces surplus value for the capitalist. This the capitalist realizes in the purchase of the commodity. But to the extent that the capitalist can persuade the worker to also waste part of the use value of the commodities he has created, the capitalist is able to produce more and sell more and so realize greater accumulation of capital. In the mid-1950s, Victor Lebow wrote in the *Journal of Retailing*, "Our enormously productive economy... demands that we make consumption our way of life, that we convert the buying and use of goods into rituals, that we seek our spiritual satisfactions, our ego satisfactions, in consumption . . . we need things consumed, burned up, worn out, replaced, and discarded at an ever increasing rate."[4] At a given standard of living, more surplus value can be realized from the worker.

Just as capitalism must use up and discard products at an ever increasing rate, as Lebow understood, so it must do the same with laborers. Those laborers are, in fact, the agricultural population that has historically been swept off the land to make way for capitalism. Once in the cities, they'll produce solid waste for the waste companies to profit from trucking back to the country. But more and more, as cheap, discardable products manufactured in East Asia fill the Wal-Marts, it is the labor of those in the Global South—the use values they create—that is being burned and buried in landfills.

It is indeed one of the most critical contradictions of capitalism that it is the useful things, with utility, with use value, that must be discarded, at the same time creating toxic nightmares, to ensure the profits and the accumulation of the capitalist system. Production, or labor, generates all value—too much value for capitalism to function smoothly. It would be a significant brake on the system if more of the use values produced were actually utilized, but the critical function of waste in capitalist society is to reduce the amount of available use value while increasing exchange value, for profits. This system enables further production whereby more surplus value accrues to the owner in the production process and is valorized in the market. As Marx observed, "the poverty of the people must be ensured," and what better way to do it than to create a system where people pay for having that which they have created and which is useful to them destroyed. It is indeed fitting that this system must also waste the creative human potential of an entire class of individuals, the largest in society. The necessary creation of waste under capitalist production is at the root of the creation of inequality and poverty, not only within a nation but globally as well.

As the problem of waste is much greater in societies where all labor tends to be exchanged through the market, the world has to look forward not to less waste throughout the world but to more as societies in the Global South "develop" and are brought into the global market in a capitalist way.

As Marx argued, no social development in capitalist society has entirely positive or entirely negative effects. For example, Marx pointed out the alienation and suffering of workers under capitalism but noted the necessary development of the technological forces that it brought. Marx believed that it was a necessary stage in the historical development of the productive forces required for the ushering in of a more democratic society. Early on, Marx conceptualized the realm of necessity and the realm of freedom. Society must produce its material needs. Some degree of economic growth would be necessary for an increasing population. But profits and accumulation were not to be the driving mechanism of society. In the realm of freedom, people could begin to produce humanly and freely because they would be free from need. The most crucial concerns of society could be addressed, then, in a truly rational way. Such a view is totally precluded by the logic and dynamic of capitalist accumulation.

Similarly, in the waste industry, there are positive and negative trends while the large companies make decisions on the basis of the bottom line. Recycling has increased when it is possible for companies to make a profit from the waste, usually as a result of public subsidies. Taken as a whole, the system militates against source reduction and the ability of local communities to establish recycling industries, as illustrated by the Supreme Court case of *Carbone v. Clarkstown*, discussed later in this chapter. Much more must be done in the area of recycling and cleaner production.

Marx noted that capitalism tends to produce both national wealth and poverty of the workers, two sides of the same coin. In the interests of national wealth, "the poverty of the people is ensured."[5] The poverty of a significant portion of the American people has most certainly been attained in America, the wealthiest nation on earth. Similarly, Robert Brenner has recently shown that the other side of the triumph of American capitalism over West Germany and Japan is the poverty and degradation of the wage worker in the U.S. economy.[6]

While it is true that the New Right dogma is firmly in place as an emerging dominant global ideology among national elites, it is still challenged from below and has little appeal, except in terms of the material results, to many at the bottom of the heap in the Global South. These fall woefully short for the vast majority living under the new global ideological hegemony and increasing global inequality. We see within capitalism an ongoing class struggle from above and below:[7] workers deprived of the use value that they produce and most unable to aspire to the realm of freedom that would allow the fulfillment of human potential, rather than a degrading culture.

Capitalist Accumulation in the Waste Industry

The waste industry experienced rapid growth in the 1980s, with somewhat slower growth in the 1990s. Indeed, the profits of the waste industry are largely an indicator of the economy itself, with growth and profits tending to fall when the economy is in a recession. According to James Cosman, president of Republic Services Group, "the more waste going into your landfills, the more profit."[8] This succinct statement suggests how the industry profits from greater consumption and waste and, in some areas, such as the high population centers of the east and west coasts, from the crisis created by where to put the waste.

The solid and hazardous waste industries are part of one of the fastest growing industry sectors in the U.S. economy and worldwide, the so-called "environmental technologies." The global market for environmental technologies was about $420 billion in 1995 and has been projected to grow to $600 billion per year by 2010. While growth in the environmental technologies industry is projected at around 4 percent per year, the U.S. industry is banking on exports to the much faster projected growth areas abroad, particularly in Latin America and Asia, where growth is projected at some 10 percent and 16 percent respectively. Solid waste management is the largest segment of the environmental industry, bringing in $34.9 billion in revenue in 1997. Hazardous waste management revenues for 1997 were $5.8 billion. The value of total services came to $111.1 billion. In the U.S., annual growth from 1991 to 1995 was 3 to 5 percent, down considerably from the late 1980s.[9]

The waste industry in the United States is a monopoly industry. It was centralized with mergers at a record pace in the 1990s. Centralization in the industry is clearly evident in the fact that the top 100 companies in the solid waste industry control some 63 percent of all solid waste revenue—$22.9 billion in 1996.[10]

Centralization of capital has come about quite rapidly, driven primarily by government regulation and capital requirements for companies to operate and carry large liabilities. Paul M. Sweezy has demonstrated that the formation of monopolies comes about through the concentration and centralization of capital.[11] Concentration occurs when individual capitalists accumulate so that the quantity of capital under each capitalist's control increases. Centralization happens when capitals already in existence are combined, as in the case of mergers. A striking example is the recent merger, in 1998, of Waste Management Incorporated (WMI), the largest in both solid and hazardous waste, with the fourth largest, USA Waste Services, Inc.

In the U.S., banks are heavily invested in the waste industry, and in recent years companies have been forced to seek credit from banks for further acquisitions of other companies and landfills—or go out of business. A second means of concentrating capital is stock companies. Most waste companies have been forced to "go public," offering their stock to the public. With stock companies, there is a large expansion in the scale of production; capital assumes the form of social enterprises.[12] There arise interlocking boards of directors with "a community of interests."

Sweezy notes that "the final stage in the development of monopoly capital comes with the formation of combinations which have the conscious goal of controlling competition."[13] When centralization has reduced the number of competitors, competition becomes cutthroat competition, and the ground is laid for a "combination movement." There is a community of interests between competitors, and common banking connections smooth the way.[14] A move is made "in which business is allocated according to a formula agreed upon among the participants."[15] With the cartel, a central committee is responsible for fixing prices and production quotas, and it has the power to punish violators by fines. This sort of arrangement was clearly seen in the "Chicago rules."[16] Sweezy points out that in the U.S., the legal restrictions on cartels and trusts have operated to increase the number of mergers, and this development increases monopoly power in the industry.[17] When a few large companies control 70 to 80 percent of an industry, "competition of a dangerous kind" is largely abolished. By 1996, in the U.S. solid waste industry, more than 60 percent of revenues, as noted above, were controlled by the top hundred companies; but there were even fewer companies in the hazardous waste sector. Monopoly control in an industry increases profits, which come essentially as a deduction from the surplus value of other capitalists or a deduction from the wages of the working class.[18] Also the rate of capitalist accumulation under monopoly capitalism tends to be higher.[19]

The amount of capital in the waste industry has expanded dramatically. Companies are faced with two choices: sell or "go out and get access to
capital to comply with the regulations . . . and adopt a growth strategy."[20]
Dale Nolder, director of marketing for Superior Services, Inc.
(Milwaukee), predicted that "in the future, 70 percent of the waste business will be in the hands of the top four or five private companies; only 20
percent will be in the hands of municipalities; and 10 percent will fall
under the jurisdiction of Mom and Pop operations."[21]

A number of mechanisms currently contribute to overall concentration and centralization of capital in the waste industry. Among these are
economies of scale, increasing privatization, subsidies and tax breaks,
transfer of liability, the Commerce Clause, emerging technologies, and
urbanization and privatization in Third World cities.

Increasing Economies of Scale

One development that is driving concentration and centralization in the
waste industry is increasing economies of scale. Big companies operating
over a wide territory, collecting and separating waste into more than 30
waste streams, and integrated with other industries that utilize portions of
the waste can handle waste more cheaply and utilize the recycled materials for other industries they own. Big companies can handle waste for up
to 40 percent less than what the process costs smaller companies.

New legislation to upgrade landfills under RCRA, Subtitle D, has contributed to monopoly, the size of companies—and fewer landfills. With
some 2,600 landfills, instead of the previous 8,000, waste is hauled further,
encouraging barging and rail haul operations. Big, publicly traded companies have largely taken over the independent haulers. Waste-by-rail and by
barge are on the rise, particularly from large population centers on the east
and west coasts, to deliver high volumes of waste to new, huge landfills. In
one such operation, Waste Management Inc. (WMI)/Atlantic Waste in
Waverly, Virginia, ships waste by Conrail, 60 cars and 240 containers a day.
WMI handles waste from the Bronx in New York City. At the Bronx transfer station, 1,600 tons per day of solid waste is compacted into 20-ton bales,

extruded into a container. Containers are shipped four to a car in 60-car unit trains from the Bronx to Atlantic Waste in Waverly, Virginia. The trains reach the landfill in 27 hours or less, stopping only for crew changes.[22]

Increasing Privatization

A second mechanism driving waste industry profits and centralization is increasing privatization by municipal governments of waste collection and disposal. Only 33 percent of solid waste today is handled by the public sector through municipalities and counties. A useful device for capitalist profits introduced by New Right approaches is "managed competition," increasingly used by local governments. Managed competition, a euphemism for privatization and contracting out, was pioneered by Phoenix Public Works 20 years ago. Such privatizations are pushed by right-wing organizations such as the Reason Public Policy Institute (RPPI) in Los Angeles. The buzzword is "competitive neutrality," which means that municipalities are forced to go head to head with monopoly waste giants to bid for contracts. Essentially, this concept forces the public sector either to give contracts to the waste industry for profits or to cut labor costs to the point that the public sector can compete with private companies. No "level playing field" really exists here because big companies spend enormous amounts on bids and have enormous economies of scale. In terms of collection, as opposed to hauling and landfilling, municipal services are still clearly large, although this picture is changing as more privatization takes hold. One indicator is that the market for U.S. municipal services comes to just $1.9 billion, while the annual revenue of Waste Management in 1996 was some $9.2 billion, with Browning Ferris reaping $5.8 billion.[23] Once the waste is collected, it is the private sector that transports it by barge, rail, and truck to distant landfills and Materials Recovery Facilities (MRFs). The future trend is for smaller municipal landfills to be closed, making way for disposal in larger, privately owned ones by the private monopoly waste industry.

A number of factors, of course, make it very difficult for the public sector to compete with big private monopoly companies. The factors driving the process include cutting costs, efficiency of economies of scale, political pressures, capital costs, Subtitle D regulations under RCRA, and the *Carbone* decision—the loss of flow control of waste by the public sector.[24] The Commerce Clause means that waste companies can take their waste anywhere, which cities and municipalities will not do. If the county or municipality owns a landfill or incinerator, it must attract enough waste to operate it economically. The Supreme Court decisions on the Commerce Clause, particularly the *Carbone v. Clarkstown* decision on flow control in 1994, means that states and local governments cannot control the flow of waste, while the waste giants have complete freedom to do so. In addition, the public sector must often provide some benefits to workers, while the private sector may not. So the public sector is forced to lower wages to the absolute minimum. Obviously, "the two sectors are becoming more and more alike," which is exactly what the thrust of "managed competition" is supposed to achieve.

Subsidies and Tax Breaks

A third contributing factor to concentration and centralization of capital in the waste industry is the provision of subsidies and tax breaks for waste companies. One example is the Landfill Methane Outreach Program (LMOP) begun in 1994. An emerging technology is landfill gas to energy plants (LFGTE). Landfill gas, methane, is a "greenhouse gas," and operators who convert the gas to energy are eligible for emissions credit trading. Methane emissions reductions have a monetary value that may be marketed to other companies or "banked" for future use. As the market emerges for these credits, companies such as Centre Financial in Chicago and Trexler & Associates in Portland, Oregon, are investing in the market. By 1998 some 170 landfills in the U.S. were converting methane to a marketable energy source.[25] There are no real emissions reductions, however, just tradeoffs, because when a company reduces its emissions of methane from a landfill by a certain volume, this allows another company to increase its emissions by an equal amount. The total volume of emissions is not reduced, so the claim that "green power" is being produced is deceptive. In addition to corporate profits for this commodity, the "green power" produced from the gas is marketed at a premium to environmentally conscious customers who are willing to pay $3 to $5 more per month. But how many consumers know that firms receive emissions credits and sell these on the market allowing them to purchase the "right to pollute"?

Transfer of Liabilities for Pollution

A fourth and extremely important mechanism is the transfer of liabilities for pollution to the public sector. With landfills, the future burden falls on the public sector once the period of monitoring is over and the landfill becomes a park, golf course, or mini–ski hill.

Still another option is simply to declare an industrial site as "clean" and provide immunity from liability to companies, under the Brownfields Economic Redevelopment Program. The Brownfields Program was started by the EPA in 1995 to redevelop abandoned or unused buildings or factory lots and buildings in urban areas. At least 40 states have participated, providing "economic incentives." Sometimes properties are capped with asphalt or cement, or the most dangerous areas are merely fenced off. These programs are designed to make "cleanups inexpensive." Here, unlike with the Commerce Clause, federalism is at work, as state agencies are given permits to draft their own "environmental cleanup methods." The idea of "encapsulating the waste in place" instead of removing it is said to be "very acceptable." One civil engineer is quoted as saying "[j]ust because it's there doesn't mean it's moving and dangerous."[26] The deeds to the brownfields properties indicate what pollutants are in the soil and protect property owners from future lawsuits over any other undetected pollution problems. More than a thousand contaminated properties in Pennsylvania had participated in this program by 1999, with at least 730 properties cleansed, if not of toxics, of legal liabilities for the owners under the "Land Recycling Act."[27]

Another development is the use of "in situ" technologies, to treat wastes on site, rather than removing them. Sometimes wastes that would normally be landfilled are used as backfill. In one case, at a former Massachusetts resort, 2,000 cubic yards of PCB-impacted soil excavated from a gun club landfill was placed in a golf course landfill prior to the construction of a landfill cap. The resort was converted into conservation land.[28]

Clearly the brownfields initiatives are designed to relieve companies of the costs of cleaning up toxic externalities. Perhaps more onerous are the recent audit immunity laws that basically protect companies from liability from infractions of environmental legislation if they audit themselves. Large waste companies also profit from immunity from liability on recyclables. They may use cheaper, more integrated recycling technologies [no prior separation] while being free from liability if materials are contaminated.

So-called remediation of toxic sites, such as Superfund sites, is another profit-making activity for corporations. Clearly in the case of a company like WMI, which owns Rust International, one branch of the company profits through the creation of Superfund landfills, while other divisions of the same company profit from remediation, largely paid as a public subsidy to the company. Across the board, the strategy of providing legal immunity to large corporations seems to be a pervasive trend.

Toxic Waste and the Commerce Clause

A fifth crucial mechanism, which cannot be overemphasized, is the use of the Commerce Clause. The three pivotal cases—*Philadelphia v. New Jersey* (1978), *Chemical Waste Mgmt., Inc., v. Hunt* (1992), *and C & A Carbone, Inc., v. Clarkstown* (1994)—have clearly had a momentous impact upon the profits and capitalist accumulation in the waste industry and are continuing to drive the centralization of capital in this industry.

Hazardous and solid waste incinerators cannot be established unless they can be sited, and the siting process itself became the greatest obstacle for industries in the 1980s and 1990s. Between 1982 and 1990, some 248 incinerator projects were canceled, due largely to citizen protest.[29] In some states, such as Alabama, state laws allowed companies allied with political and industry groups to put a hazardous waste facility in without the knowledge of the local community. The prime example is the largest toxic waste facility in the United States, located in Emelle, Alabama.[30] At first, the state supported the waste facility. When the state later tried to limit the import of PCBs by Chem Waste Management, it ran up against the power of the company to use the Commerce Clause to preempt local control.

In 1991, the Emelle facility, operated by Chem Waste Management, Inc., a subsidiary of waste giant Waste Management Incorporated (WMI), was one of only eight in the U.S. permitted to dispose of PCBs. In 1989, the facility handled 788,000 tons, 8.6 percent from in-state and about 17 percent of all hazardous waste commercially landfilled in the U.S. Some 40,000 truck loads were dumped on 300 acres.[31] Dumping in Sumter County follows the familiar pattern where people are likely to be poor and black.[32] With no public hearing, most people either knew nothing of the facility or thought it was a brick, cement, or fertilizer factory. The Chem Waste plant had 450 employees in 1987 and a payroll of $9 million, with 1989 user fees of $3.64 million. Many civic organizations in the county were given a cut of the pie. Two explosions occurred at the plant in 1981, after which the workers protested working conditions.[33] Two environmental organizations emerged. Sumter Countians Organized for the Protection of the Environment (SCOPE), which was largely white and moderate, claimed to want better monitoring and information. Alabamians for a Clean Environment (ACE), more radical and also largely white, had a membership of some 300, including a core group of around 10. ACE was trying to close the plant. ACE was seen as "utopian," a "mouse trying to stomp an elephant." But ACE linked to national organizational networks, such as the Sierra Club, Greenpeace, and Citizens' Clearinghouse for Hazardous Waste. In 1981, a black organization called the Minority People's Council organized a demonstration at the plant gates protesting unsafe working conditions.[34] In the 1987 demonstration, ACE, Greenpeace, and Citizens' Clearinghouse launched a demonstration of the "Toxic Trail of Tears" at "the nation's pay toilet."[35]

One of the local newspapers, the *York Sumter County Journal*, supported by the company, "jeered" the environmentalists in weekly editorials in a section called "Cheers and Jeers" and refused to print letters to the editor. The position of the Alabama paper was that Chem Waste

Management paid for space and that environmentalists such as ACE were free to do the same. It was only with the Texas PCB suit that the *Journal* decided to go against the waste company. But by then it was too late.[36]

In 1988 the state of Alabama, through the governor's office, filed a lawsuit attempting to keep 47,000 tons of soils contaminated with PCBs from Texas out of the state, after Chem Waste acquired a permit to dump them at Emelle. The Alabama Department of Environmental Management (ADEM) had earlier supported the company but switched their policy after the Texas PCB issue.[37] But at this point it was too late. Armed with the heavy artillery of the Commerce Clause, Chem Waste swung its big guns directly toward the capitol and the governor, and company officials geared up to pulverize anything that came in their way.

Opposition peaked with the PCB issue and galvanized opinion across the state. The Alabama legislature had imposed a fee of $72 per ton on out-of-state hazardous waste and a $25.60 fee on all waste regardless of origin. Legislators also attempted to place a cap on the amount dumped at Emelle between July 15, 1990, and July 14, 1991. The governor, attorney general, and head of ADEM sued as private citizens in the case of *Chemical Waste Management. Inc. v. Hunt* (1992).[38]

There is a significant history of cases dealing with the Commerce Clause. Article I, Section 8, makes clear that Congress has the power to regulate interstate commerce. Hazardous waste has been treated by the courts as an article in interstate commerce.[39] The Eleventh Circuit of Appeals held that the plaintiffs lacked standing to sue and that the state had failed to establish a causal connection between the alleged injury to Alabama's environment and the lack of notice and opportunity to participate in the selection of the remedial action.[40]

Following this defeat in court, a new bill was passed in the Alabama state legislature, the Holley Bill, which banned waste from certain states that did not have their own hazardous waste facilities and had not entered into the interstate or regional waste agreement. Once again the Holley Bill collapsed, battered by the Commerce Clause, as an unconstitutional barrier to the movement of these wastes, even though they were PCBs.[41]

Alabama followed up with another law involving an additional fee on out-of-state waste, a base fee plus a cap on wastes. Chem Waste filed for *certiorari*. The Court agreed to decide the issue of the additional fee. On June 1, 1992, the additional fee was held to be unconstitutional, as well, under the Commerce Clause.[42] The Supreme Court stated that "[n]o state may attempt to isolate itself from a problem common to the several states by raising barriers to the free flow of interstate trade."[43] According to the Court, Alabama had not "demonstrated that the discrimination against out of state waste is based on a valid factor unrelated to economic protectionism."[44] The PCBs enjoyed the constitutional freedom of movement. Those at the helm of the state did not realize their powerlessness in the face of a major corporation until they were rebuffed repeatedly in court at the hands of the Court's interpretation of the Commerce Clause.

Similar cases have occurred in Michigan and Oregon. Michigan failed in its attempt to keep 1,750 tons per day of out-of-state solid waste going into a landfill. The Court said a county could not isolate itself from the national economy. Michigan's ban on the waste, according to the Court, could not relate to health and safety concerns, as it did not apply to locally generated wastes.[45]

In another Commerce Clause case, Oregon tried to impose an out-of-state surcharge on solid waste. The Court found this was economic protectionism and hence unconstitutional under the Commerce Clause. Oregon argued that its action was protective ("resource protectionism"). Here the Court ruled that a state may not accord its own inhabitants a preferred right of access over consumers in other states to natural resources located within its borders.[46]

The most significant recent case is that of *Carbone v. Clarkstown*, which has empowered the waste industry even more by outlawing flow control. The Supreme Court ruled that a flow control ordinance requiring all non-hazardous solid waste within a New York town to be deposited at a transfer station owned by the city, but operated by a private contractor in the town, violated the Commerce Clause. The waste industry correctly interpreted this decision as giving waste companies much greater freedom in the handling of waste, and the ruling has contributed significantly to the growing size of waste companies and centralization of capital in the industry.

The implication of these cases is that once a hazardous waste facility is established, local communities and even states will probably not have the legal power to decide what will go into a landfill nor how much. That will be decided elsewhere, either by private waste corporations or—often in the case of hazardous waste—by the courts.

Cement Kilns and Industrial Boilers

New technologies are emerging, such as incorporating hazardous waste ash into glass and asphalt and using contaminated soil in asphalt. Also, the use of liquid hazardous waste for industrial fuels for boilers and cement kilns is widespread and increasing.[47] Cement kilns burning hazardous waste are de facto hazardous waste incinerators and produce dioxin (2,3,7,8-tetrachlorodibenzo(p)dioxin), the most dangerous of the 75 forms of dioxin. This type of dioxin ends up in the food chain in milk, eggs, and meats. The industry argues that cement kilns are "superior technology" for the management of hazardous wastes.[48]

Today, Portland cement kilns are viewed by the cement industry to be "effective means of recovering value from waste materials."[49] Cement production is energy intensive and involves heating calcium, silicon, aluminum, and iron oxides to over 2,600 degrees Fahrenheit in furnaces to produce cement clinker, which is then ground into cement. Technicians argue that there is almost complete destruction of organic constituents in fuels containing hazardous waste. But many stacks of cement kilns are said to have a destruction and removal efficiency (DRE) lower than 99.99 percent, and a DRE as low as 95 percent has been documented.[50] One consultant quite explicitly points out that "[s]cientific issues concerning emissions and potential effects on health and the environment are necessarily secondary to economic and political issues in the United States."[51] Hence, more and more hazardous waste is being recycled into usable products said to be safe, but often the process of production involves the incineration of hazardous waste in a facility disguised by a different name.

Urbanization and Privatization in Third World Cities

The largest companies, like WTI, are also increasing concentration and centralization of capital as a result of urbanization and privatization in large Third World cities. This trend is part of the continued enclosure of the land, driving peasants into the cities as a result of the pattern of development—a continuation of the historical "enclosure" process that Marx described in *Capital*. Here, uneven development creates a significant opportunity for waste companies to make large profits from rapid urbanization and the expansion of the consumer revolution among the emerging middle classes. Sweeping the agricultural population off the land and into the cities not only makes room for the waste but also creates more waste. The increasing hegemony of New Right ideologies of absolute property rights are tailor-made for the needs of the waste industry. Robert Nozick's *Ideology, State and Utopia*, which argued that the rights of property could never be limited for the public good, has surely had considerable impact.[52]

Conclusion

The global political economy today functions in an iron framework put into place largely by the United States, in coordination with the states of Western Europe, since World War II. Today one of the central institutions of that framework is the U.S. dollar, increasingly the global currency. The value of the dollar and the power of U.S. transnational corporations are backed up by the International Monetary Fund, the World Bank, and the International Trade Organization—all backed by the political and military power of the United States—for the fundamental purpose of corporate profits and capitalist accumulation. These organizations have increasingly functioned as levers to force every nation into opening its markets to U.S. capitalist investment and profits. Global resources, including human capital in the form of labor, are to be readily available to transnational corporations, wherever they are located, in the so-called New World Order. Where such resources are not readily available in the market, the U.S. military, the gendarme of U.S. transnational corporations, can be brought in to soften up the resistance. Force and this global institutional structure are used globally to neutralize attempts of any nation to practice sovereignty over their own resources and domestic economic policies. This is what integration of the global markets is all about. Similarly, when small communities in the U.S. wish to exercise local autonomy in the use of their own resources, they run up against the increasing power of the global corporation. New Right ideology, with its sanctification of the absolute rights of private property in the hands of gigantic global corporations, is a powerful hegemonic force. The capitalist state–global corporate power conglomerate constitutes a voracious predator of the environment and ecosystems wherever they are located.

The underlying logic of the capitalist system necessitates the production of toxics and their necessary release into the environment to sustain profits and accumulation in the competitive global marketplace. The public sector guarantees and ensures profits, in part, by paying the "toxic debt." There are no solutions, no policies that can be forthcoming, as long as the capitalist logic prevails. Even recycling, often touted by environmentalists as the right solution, is an integral part of the monopoly capitalist waste industry today. While it does tend to keep a proportion of recoverable materials out of landfills, it is not, by any means, the answer to the waste problem. Recycling tends to shift the responsibility for the management of waste from corporate producers to the individual consumer. Moreover, recycling preserves valuable landfill space owned by large waste companies. Landfill space is a valuable commodity that enables a company to reap a high rate of profit because of its scarcity under new regulations. This can be seen in California, where the Integrated Waste Management Act of 1989 required cities to divert 25 percent of their waste streams by 1995 and 50 percent by 2000. The California Integrated Waste Management Board (CIWMB) reported in 1998 that a million tons of waste had been diverted from landfills in California between 1990 and 1996. The impact was to free up landfill space so that California now has some 28 years of landfill capacity.[53] Recycling and new products are often publicly subsidized while the larger profits are made by the continued dumping in the freed-up landfills. Finally, the overall impact is not very great, in terms of the total of volume of waste produced and disposed of in the United States.

The needs of major corporations for continued capitalist accumulation can be expected to prevail. While policies may be marginally better or worse, there is no real solution to the reality of continued environmental degradation under actually existing capitalism. Environmental policy making is largely tinkering, and efforts to get tough, as in the Superfund program, quickly get eroded when they impact the ability of firms to continue the historical rate of capitalist accumulation. Particularly under the tendencies of New Right agendas that further hollow out the authority of national and local governments and further weaken regulation for the public good, corporations increase their hegemony over civil society. The New Right neoliberal agenda being applied to countries around the world is all about weakening the ability of the state apparatus to regulate transnational capital, weakening control of national governments, and rendering individuals powerless under the rubric of "liberal democracy."

In the United States, local governments are stripped of autonomy, to the greatest extent possible, under the principle of preemption. By 1998, some 24 states had passed "corporate secrecy" laws, also known as "environmental audit privilege" laws.[54] Under these laws, companies typically "audit" themselves and must promptly report any infractions of the rules to government officials; if "reasonable" efforts are then made to come into compliance, they are exempt from any fines, and the documents are secret and cannot be revealed to the public or used against them in litigation. Those who do the audit do not have to testify in court. These laws would seem to effectively annihilate any "bad-boy" legislation since declaring documents secret can also be used retroactively. A few states have laws that include anti-whistle blower clauses. In Texas, if an employee or government official reports an incident to the public and the company is fined as a result, the whistle blower is held legally responsible for the fine and other penalties. This law would make it extremely difficult to enforce any environmental legislation, which is, of course, the very purpose of such clauses. In 1997, then Senate Majority Leader Trent Lott sponsored an audit privilege bill at the national level. For Peter Montague, such laws amount to a "right to know nothing," and they reveal "just another aspect of the rapidly growing power of corporations in America and worldwide."[55] At the international level, the International Monetary Fund (IMF), World Bank (WB), World Trade Organization (WTO), as well as emerging trade agreements such as the North American Free Trade Agreement (NAFTA) serve, in an analogous manner, to render relatively powerless any governments that would act to protect their own resources and markets from the needs of global corporations for continued capitalist accumulation. Thus the continued hegemony of the global capitalist economy under "globalization" is clearly linked with internal hegemony over resources in local communities. The continued process of enclosure and accumulation of resources in the global periphery and the domestic periphery enhance the bottom lines of major global corporations. The government levers of the state apparatus exist to ensure access of big corporations to the resources needed for accumulation and to discipline the unruly "underclasses," the majority of the people. As accumulation is more exploitative in the "developing world," state brutality in hand with transnational capital is often more unfettered there.[56] This is part of the neoliberal global agenda.

The loss of power by people in local communities, and around the world, has led to a movement with more than 5,000 anti-toxics groups struggling for environmental justice and democracy in the U.S. This struggle involves the power of people to control their health and the environment in their neighborhoods.[57] A global network of environmental organizations is emerging through which people demand justice and grassroots democratic control. As such, they are forced to confront both intransigent governments and the increasingly global hegemonic forces of the market underwritten by corporate propaganda.

3
Environmental Justice, Democracy, and Grassroots Political Struggle

The lesson of the Movement is that we are the power.... This struggle is not just about environment but about basic issues of justice and fairness, of right and wrong, of the have-nots and those with the economic power who would seek to exploit all of us.

— Penny Newman[1]

Capitalist accumulation in the toxics economy requires that many communities become sacrifice zones for the dumping of the enormous amounts of toxics produced through production techniques that enhance profits. Clearly, environmental policies of governments and corporations have targeted the weak in American society. This is quite evident in the frequently cited 1984 report of Cerrell Associates to the California Waste Management Board. In the report, entitled *Political Difficulties Facing Waste-to-Energy Conversion Plant Siting*, the conclusion reads: "All socioeconomic groupings tend to resent the nearby siting of major facilities, but middle and upper socioeconomic status possess better resources to effectuate their opposition. Middle and higher socioeconomic strata neighborhoods should not fall within the one-mile and five-mile radius of the proposed site." Target communities should include those under 25,000 population, be rural, politically conservative, free market orientated, above middle age, with high school or less education, farmers, miners, those with low income, and those who are likely to see significant economic benefits in the waste industry for the local community.[2]

The environmental justice movement emerged in the 1990s in the United States as part of the grassroots environmental struggle by community activists fighting any number of toxic battles. The term "environmental justice" itself emerged from the work of antitoxics activists, especially in the South. The EJM linked the issue of social, economic, and political marginalization of women, minorities, and poor communities to environmental issues such as toxic pollution, the right to a clean environment, and workplace hazards. The EJM can be traced back to the 1960s Civil Rights Movement, urban disturbances over uncollected garbage, and to the Memphis sanitation workers' strike in 1968. This movement, with over 7,000 community grassroots groups nationwide,[3] has expanded grassroots democratic environmentalism in minority and economically depressed communities and raised the awareness of environmental racism and class discrimination.

The major issues of the EJM include toxics, waste facility siting, urban industrial pollution, childhood lead poisoning, farm workers and pesticides, land rights, sustainable development, export of toxics to the Global South, risky technology, and leaking landfills (which include virtually all landfills).[4] The EJM has also led to demands for a superfund for workers, restrictions on capital flight, elimination of production of toxic substances, less-polluting transportation systems, environmentally sound economic development, equitable distribution of cleanup programs, international laws that protect the environment and worker rights, and justice for the poor in countries of the Global South.[5]

The environmental justice movement is a broad, comprehensive movement for social justice, organized around professional networks of organizations, such as the Citizens Clearinghouse on Hazardous Waste (later changed to the Center for Health, Environment and Justice), the African American Environmental Association, the Southwest Organizing Project, the Southwest Network for Environmental and Economic Justice, the Southern Organizing Committee for Economic and Social Justice, the Gulf Coast Tenants Association, the African-American Black and Green Tendency in California, and many others. It is multiracial and multiethnic, uniting African Americans, other ethnic groups, women, and low-income communities around environmental justice issues.[6] Robert D.

Bullard points out that the EJM is characterized by grassroots activism and is a bottom-up movement. This suggests that minority communities have always been interested in environmental issues but were previously excluded from the mainstream movement.[7] These grassroots groups serve as information centers for the assistance of local environmental groups all across the country.

Environmental justice groups often emerge from organization and mobilization around a single issue and move on to multi-issue agendas as they network with other groups. Local groups conduct their own research in their own communities to challenge the experts and demand roles in decision making about issues which directly affect their lives. Through active involvement in such groups, people of color, women, and members of working-class communities gain opportunities to develop leadership skills and to share their experiences with other grassroots groups.

The environmental justice movement has not only brought African Americans, Mexican Americans, Native Americans, women, and low-income white individuals into the environmental movement, it has succeeded in putting the agenda for environmental justice on the national public agenda. After a series of national studies and conferences, such as the First National People of Color Environmental Leadership Conference in 1991 in Washington, D.C.,[8] and conferences at the University of Michigan School of Natural Resources, activists made demands directly to national officials for action. In 1992, the EPA created the Office of Environmental Justice, and on February 11, 1993, President Clinton signed Executive Order 12898, which directs federal agencies to incorporate environmental justice principles as part of day-to-day operations. Grassroots groups work to hold the government accountable for policy actions and demand policies that actually result in the improvement of the environmental health of communities and protect the people who live there. Mainstream environmental groups, such as the Sierra Club, the Wilderness Society, and the Conservation Foundation, were pressured to diversify their leadership from their mostly white male staffs.[9]

The EJM embraces both liberal reform, at one end of the spectrum, and radical demands for an alternative economic system at the other.[10] Radical groups may challenge the fundamental social, economic, and political structures and institutions of the U.S. and global political economy, which are believed to contribute to inequality and injustice for minority and low-income communities. Such inequality is evident in toxic pollution of air and water, increased sickness and disease, increased incidence of cancer, deteriorating communities, and unsafe workplaces for the politically weak. For radical environmentalists, the quest for development, even if viewed as "sustainable development" and "progress," puts a disproportionate share of the costs on minority and low-income communities and is at the root of the present national and global environmental crisis. This situation can only worsen if corporate exploitation of the earth's resources is not challenged by a different pattern of development of communities that puts people before profits, empowers people to make decisions that protect their livelihood, and ensures a healthy environment and future for the children of all groups and classes in society.[11]

At the beginning of the new century, one of the most pressing issues continues to be waste, both in the form of household solid waste and hazardous waste. EJM struggles continue to revolve around solid and hazardous waste landfills, solid and hazardous waste incineration, the increasing use of hazardous wastes as fuels for cement kilns and other industrial purposes, the rail and barge shipments of solid waste for hundreds of miles into local communities, and other environmental inequities.[12] Atmospheric toxics and cleanup of old industrial sites in urban slums are salient issues for both big corporations and EJM activists.[13]

The right to know the environmental record of companies is also an issue under resistance from corporations.[14] Corporate secrecy laws suggest that the power of monopoly capital is a greater threat than ever to people struggling to hold onto their democratic rights. Various means of hollowing out the power of local governments, under the concept of "preemption," have made EJM struggles more and more problematic.

The Dynamics of EJM Struggles

Celene Krauss has attempted to understand the participation of blue-collar women in the EJM in terms of ideas from feminist writings and new social historiography. Krauss argues that these approaches provide theoretical perspectives that are helpful in understanding the emergence of grassroots environmental political struggles, more broadly. Five key mechanisms are at work: First, there is the emergence of a critical political consciousness in practice, or, more specifically, in confrontation with the industry, state, and political power complex. Second, as a result, there is increasing understanding of the link between private encroachment and public power. Third, the emergence of a political struggle grounded in popular culture and the perception that deeply held traditions and values are being violated. Fourth, the loss of community and livelihood with the encroaching intervention of the state and the market. Fifth, the emergence of a strong leadership role by marginalized peoples such as minorities, women, Native Americans, and the poor. Women are concerned about the health of their children as well as the broader community.[15]

The dynamics of EJM struggles certainly bear out the profile Krauss offers here. The weakest sections of the society confront some of the most powerful forces, such as huge corporations and governments and their flocks of highly paid, high-flying lawyers. These struggles tend to have a David and Goliath character about them. Environmentalists are forced to depend upon free legal defense from lawyers who sympathize with their cause. State intervention may become the "kiss of death."[16] On the other hand, the state can sometimes assist, when environmentalists force the state to obey its own laws.[17] The state may play a positive role in environmental struggles, for example, in closing landfills that are polluting the groundwater and shutting down waste incinerators that are threatening people's health. More often than not, however, the political, economic, and legal structures militate against environmental and health concerns. Politicians know they cannot win elections without the campaign money from the most polluting corporations in their states. Economic growth and corporate profits tend to be the bottom line.

When confronted with such a challenge, people develop a critical consciousness. This insight begins to "unmask" the ideology of popular participation and democracy. Political consciousness arises and is continuously molded from the everyday world of experience.[18] People expect the political process to work for them, assuming that their elected public officials are on their side. As Lois Gibbs wrote, "I believed in democracy, but then I discovered that it was government and industry that abused my rights. But my experience is not unique."[19] When people discover that government and industry are working against their interests, which happens often in environmental struggles, their political consciousness is heightened. Their struggle also "unmasks" the ideology that hides the relationship between their private lives and public policies.[20] In environmental struggles, those at the margins of society usually find that they are expendable, to be made sacrifices for the more powerful, the industrial culture, and "progress," under the guise of "economic development."[21]

Grassroots struggle is a process whereby those on the periphery of the society and economy confront those at the center, the owners and decision makers. Those on the periphery tend to be women, minorities, rural residents, Native Americans, and the poorest segments of society. In other words, those marginalized politically and economically. Since these marginalized elements—the average person, in Walter Lippmann's words—are supposed to listen to the managerial class in American society,[22] their entry into the political process is often seen as deeply subversive of the prevailing and established political process in the United States. It is. They challenge business as usual, whereby the corporations run the economy, and business is carried out through the relatively controlled process of elections, courts, legislation, and so on. For example, it is clear that in many cases, public hearings on environmental matters are not meant to affect the procedure of approval or disapproval of projects; this process primarily serves to allow the public to let off steam.[23] However, the system operates within an ideology that claims that democracy exists, but this belief has not actually been put to the litmus test for most people until they are involved in a struggle that directly threatens their health, lifestyle, and community.

As struggles emerge, the movement grows out of popular culture, traditions and community. This process has been called "cultural activism."[24] The debate is very much alive as to what elements such movements are rooted in. Tradition, local economics, culture, religious traditions, ethnic values, local customs, family and individual circumstances all may play a role. Local cultures differ widely even within the same state in the United States, and within different regions of the community. To some extent, it may be a matter of how dominant the state and market have become. Human activity in the periphery may be driven less by market incentives and other pecuniary factors.[25] In rural cultures, farmers may come to see a link between ecology and their rural farming culture.[26]

In states like Missouri, the rural culture differs from that of the town business elites and big business. There is a general populist distrust of big business; farmers are peripheral to the larger economy and often exploited by big business and the markets, as the producers of primary products. The rural community takes on an organic, rather than completely individualistic, character, with the continuing tradition of neighbors helping each other in times of difficulty. If a farmer falls ill, others often pitch in and harvest the crops. They often cooperate in their work, a tradition going back to the days when a work crew was required to operate the threshing machinery at harvest. When some farmers begin to be bought off by the encroaching company, it is a threat to these traditional cultural values. E. F. Schumacher's observation about the English townsman helps to clarify a difference in worldview between urban and rural thinking: The "philosophy of the townsman, alienated from living nature, who promotes his own scale of priorities by arguing in economic terms," sees agriculture merely as a "declining enterprise," and not worth preserving.[27] At the heart of the rural culture, however, is a strong belief, or at least sense, that farming *is* worth preserving; the rhythms of nature themselves have intrinsic worth, almost sacramental value.

Religious traditions also contribute to this rural culture. The popular religion encourages the ideal that one should do what is right for the land, the community, the family, and future generations, looking beyond the concern for short-term profits. Rural residents have a natural-based feeling for "intergenerational equity." A natural "land ethic" emerges from those who spend their lives working the soil. Farmers are used to bad years and good and tend to look at things in the longer term. Something that threatens to end this way of life decisively is a serious threat. The agrarian ideal, the view that farming and the rural way of life is a good in itself, remains strong. As owners of the land for a number of generations, in many cases, there is a long-term commitment to the community.

Women are particularly important in EJM struggles because beyond the influence of the traditional culture, women tend to be more willing to challenge the system when it threatens their children's health. It has been noted that 70 to 80 percent of the local leaders who challenge the system are women.[28] They are often seen as "emotional housewives" but actually "derive power from their emotionality."[29] In EJM struggles, especially toxic struggles, women generally participate more than men.[30]

Indeed, women have played a disproportionately important role in the EJM. Scholars have noted that "[W]omen have been at the forefront of every historical and political movement to reclaim the earth."[31] Housewives with no previous leadership experience have taken the lead in fighting toxics in the United States.[32] Many have become nationally known figures, among them Crystal Eastman, Alice Hamilton, and Florence Kelley, who pioneered research on workplace hazards.[33] Rachel Carson, Lois Gibbs, Penny Newman, Kaye Kiker, Cora Tucker, and Helen Caldicott are among the leading activists against hazardous wastes in their local communities and are also nationally known among environmental activists in the United States. In East Livermore, Ohio, local housewife Terri Swearingen led the fight against a hazardous waste incinerator; the struggle still continues while the incinerator burns. In Jacksonville, Arkansas, Patty Frase organized the Arkansas Clean-up Alliance. Both of her parents died after the Vertac Chemical plant burned TCDD dioxin.[34] A Louisiana woman, Emelda West, from Convent, Louisiana, traveled to Japan to the headquarters of Shin-Etsu Chemical Company to fight against the building of a chemical plant that would emit up to 450,000 pounds of toxic fumes a year.[35] These women have emerged as key figures in their communities to lead the fight against existing and proposed facilities. They have organized their communities and networked across the nation sharing information and tactics. They have become political leaders too in demanding changes in the law and the system. As one Missouri woman put it, they "get in the face" of those who control the system.

When women take the bull by the horns, this assertiveness "helps define a new source of community."[36] Antitoxics movements often become community development movements whereby new community leaders, often women, emerge. In this way, the issue of who should be leaders in the movement and in the community is raised and dealt with.[37] In such movements, women do not merely participate; they lead, often pulling in the men too.

African Americans and Other Minorities

Racial inequality among communities in terms of environmental pollution and quality of life is well documented in the environmental justice literature.[38] According to Bunyan Bryant, environmental racism "refers to those institutional rules, regulations and policies of government or corporate decisions that deliberately target certain communities for least desirable land uses, resulting in the disproportionate exposure of toxic and hazardous waste on communities based upon certain prescribed biological characteristics. Environmental racism is the unequal protection against toxic and hazardous waste exposure and the systematic exclusion of people of color from environmental decisions affecting communities."[39]

A U.S. General Accounting Office (GAO) report in 1983 stated that of hazardous waste landfill communities in the U.S., three-quarters are poor, African American, or Latino-American. The *National Law Journal* staff reported that "penalties under hazardous waste laws at sites having the greatest white population were about 500 percent higher than the penalties at sites with the greatest minority population, averaging $335,566 for white areas, compared to $55,318 for minority areas. The disparity under the toxic waste law occurs by race alone, not income. The average penalty in areas with the lowest median incomes is $113,491, 3 percent more than the $109,606 average penalty in areas with the highest median incomes. For all the federal environmental laws aimed at protecting citizens from air, water, and waste pollution, penalties in white communities were 46 percent higher than in minority communities. Under the giant Superfund cleanup program, abandoned hazardous waste sites in minority areas take 20 percent longer to be placed on the national priority list than those in white areas. In more than half of the 10 autonomous regions that administer EPA programs around the country, action on cleanup at Superfund sites begins from 12 percent to 42 percent later at minority sites than at white sites. At the minority sites, the EPA chooses 'containment,' the capping or walling off of a hazardous dump site, 7 percent more frequently than the cleanup method preferred under the law of permanent 'treatment,' to eliminate the waste or rid it of its toxins. At white sites, the EPA orders treatment 22 percent more often than containment."[40]

The Federal Agency for Toxic Substances Registry found that for families earning less than $6,000 annually, 68 percent of African American children had lead poisoning, compared with 36 percent for white children. In families with an annual income of over $15,000, more than 38 percent of African American children suffer from lead poisoning, compared with 12 percent of white children. In Los Angeles, over 71 percent of African Americans and 50 percent of Latino Americans live in communities within the most polluted area, compared with only 34 percent of whites in such polluted communities.[41]

A well-known study by the Commission for Racial Justice found race to be the most important factor in the location of toxic waste sites. Three out of five African Americans live in communities with abandoned toxic-waste sites. Most hazardous waste incineration takes place in minority communities. There are many other findings that document this racial inequality.[42]

Beverly Hendrix Wright and Robert D. Bullard also found large discrepancies between whites and blacks in occupational hazards. There is a significant disparity in life expectancy between whites and blacks, 69.6

years for blacks compared to 75.2 years for whites. Injury or illness on the job in the United States is 37 percent higher among blacks.[43] This racial discrimination is perpetuated under the myth that blacks can stand hotter temperatures in the work environment than whites. As one writer noted, "A job should not be a death sentence."[44]

Perhaps no other area demonstrates environmental racism in operation more than the 85-mile-long Cancer Alley in southern Louisiana, along the Mississippi River from Baton Rouge to New Orleans.[45] This is an example of a sacrifice zone where a colonial mentality prevails and local governments and big business take advantage of the powerless. This area has roots in the plantation system and white resistance to equal justice. States have regressive "look the other way environmental policies" and generous tax breaks to attract polluting industries. Both human and natural resources are bought by big corporations at bargain prices.[46]

Louisiana is a hazardous waste importer state that ranks 49th on the Green Index of environmental quality and 50th in toxic release to surface water, in high-risk cancer facilities, and in per capita injection wells and oil spills.

Internal Colonialism and Native Americans

An important development in the EJM is the widespread use of Native American Treaty Rights for environmental preservation and conservation of resources. These aspects of the EJM obviously have importance for tribal rights of groups in countries around the world. Some 370 treaties were made between the United States and Native American nations, whereby over two billion acres of land changed hands. By 1868, less than 140 million acres remained in Indian hands.[47] Through various treaties, the U.S. gained the rights to most timber and minerals on Indian lands. In Wisconsin these treaty rights have been challenged. An 1837 treaty that specified that the Chippewa (Anishinabe) people had the right to hunt, fish, and gather on the lands they had ceded has been tested in federal courts.[48] In Wisconsin, a case that resulted in the *Voigt* decision (1983), reversing an earlier opinion, established that these treaties have standing in law. The decision affirmed the right of the Chippewa to hunt, fish, and gather on public lands off the reservation.[49] The significance of this and similar opinions is that Native Americans have "an internationally recognized status that can be used to protect the environment, for everyone and all human beings."[50] The treaties are matters of international law.[51] Consequently, "treaty rights within the territories that were ceded are the best potential environmental protection device that currently exists."[52] According to Russ Busch, an attorney for Native Americans, "Without the treaty case law more than half our arguments against nuclear plants would have no teeth."[53]

Such treaty law was tested in the early 1980s in the case of Project ELF (extremely low frequency). The U.S. government attempted to locate an underground antenna system for the Navy in the northern Wisconsin forests for the purpose of contacting nuclear first-strike submarines. The government also attempted to locate a nuclear depository in northern Wisconsin. The struggle against these facilities involved the Chippewa along with some twenty-five Midwest tribes and activists in the American Indian Movement (AIM).[54]

Another environmental threat to the area was brought by oil, gas, and mineral companies. American Indian reservations are often sites of large reserves of minerals crucial to the U.S. economy. These include one-third of the U.S. low-sulfur coal, 20 percent of U.S. reserves of oil and natural gas, one-half of uranium deposits, and bauxite, zeolites, and also water resources.[55] There are some 40 multinational companies that lease over 400,000 acres in northern Wisconsin.[56] Local activists point out that "Chippewa treaty rights have stood as potentially the greatest obstacle to mining in Northern Wisconsin"[57] and to the sacrifice of such land for the mineral needs of the military, industry, and urban consumers.[58] Amoco Minerals is a top corporate leaseholder in northern Wisconsin for copper, zinc, lead, uranium, and thorium.[59] Rick Whaley and Walter Bresette conclude, "The willingness to extend treaties to environmental protection of all communities within the province of the treaties is a potentially powerful strategy to forward-looking earth politics."[60]

Indian lands are exploited by the uranium industry for both the military and private industry. Joni Seager has noted that "whether at peace or at war, militaries are the biggest threat to the environmental welfare of the planet."[61] The U.S. approach to dealing with tribal Indian lands has resulted in "internal colonialism" and dependency of tribal peoples on large corporations.[62] Contributing to this system was the Indian Reorganization Act of 1934, whereby a system of tribal council governments on reservations replaced traditional Indian forums. These less resistant bodies allowed U.S. development policies to go forward through the Bureau of Indian Affairs (BIA) bureaucracy. For example, the Navajo Tribal Council approved uranium mineral extraction in 1952 with the Kerr-McGee Corporation.[63]

At the Churchrock site, Kerr-McGee left behind large deposits of uranium tailings that resulted in much radiation-induced lung cancer. Radiation levels in the mine were said to be 90 times the permissible levels.[64] In the much-noted Churchrock accident in 1979, a dam broke at a uranium mill site, releasing 100 million gallons of radioactive water into the Rio Puerco River, killing 1,000 head of livestock.[65]

The American capitalist economy, which colonizes Native Americans, also colonizes other tribal communities in remote areas of the globe— ones peripheral to the state and the market. This conquest-colonial system cannot be sustained over time. The remaining resources on Indian lands and the lands themselves are targeted for extraction and for the depositing of hazardous wastes. American economic conquest has destroyed the Native American socioeconomic systems that were long sustainable.[66]

Achieving Environmental Justice: A Radical Perspective

At its most radical, environmental justice challenges the systemic structure of developed capitalist society and the global system of discrimination. John Bellamy Foster grounds global ecological decline in "the dominant conception of human freedom," which sees freedom as the selfish exploitation of nature for individual gain. The EJM often questions the notion that the "free market" is the best instrument available for combating environmental problems. Many social thinkers have seen the need for "organic community" and institutions that promote a "sense of community."[67]

Ultimately, "achieving environmental justice demands major restructuring of the entire social order."[68] Such restructuring would include a challenge to absolute property rights; a challenge to the logic of growth without limit; the right of everyone to a clean environment; the concept of security as a sustainable ecological system, rather than military superiority; and social planning and grassroots democracy as the basis for environmentally sound growth. There must be a shift of power from corporate public policy making to local policy making by the people. After all, heads of corporations are not elected by the people. Nevertheless, they run totalitarian organizations that affect thousands and even millions of people's lives. The New Right ideology, from ideologues such as Frederick Hayek, which sanctifies and deifies private property, effectively serves as the groundwork for the emerging dictatorship of the corporatariat. Francis Fukuyama has it wrong. It is not, in fact, individualism and freedom that are emerging, but a new form of totalitarian organization of society, cloaked in liberal trappings and clever buzz terms.[69] The EJM often involves a major shift in the concept of democracy as it has existed in American society—from traditional elite to grassroots democracy. People must have the right to know about health effects of toxics, the right to inspect facilities, and the right to negotiate agreements with responsible parties.

There is a clear and recognized direct conflict between environmental protection and corporate profits.[70] When the costs of capital are increased by environmental protection, states and transnational corporations seek to dilute or remove such protection through free trade rules and arrangements. Cost benefit analysis, risk management, and a narrow focus on legislative tinkering must also be challenged. The new corporate campaign of greenwashing environmentalism, backed by corporate big bucks, is another formidable challenge to the EJM. Other factors include the white-male power structure and dominant culture, the prevailing more-is-better consumerist mentality, and the deification of advertising as "the right to choose" by New Right ideologues. The challenge will involve linking grass-roots organizations to international organizations and grounding research in empirical facts. When states do not lead, the people must take action.

The EJM has moved to link academic researchers with local communities and activists through "communiversities." Gaining control of local political processes and building coalitions of grassroots organizations could push governments in the direction of new legislation and policies on hazardous waste production, disposal, cleanup, and waste facility siting. The EPA and other federal agencies must be challenged to help address and eradicate unequal environmental protection. Participation and organization can lead to local political empowerment.[71] In the words of Robert Bullard, the "ultimate goal" should be to "democratize environmental decision making and empower disenfranchised people to speak out and act for themselves."[72] Globally the movement may borrow ideas from the West while being grounded in local political cultures.

The logic of capitalist accumulation requires corporate invasions into local communities for resources. As the power of the global corporation grows in the age of neoliberal ideological hegemony and the deification of the rights of private property, people must struggle in community, and solidarity, if the earth is to be saved. Truly, the environmental justice movement must transcend race and national identities.

4
The People's Struggle Against Amoco Waste-Tech in Mercer County, Missouri

We are fighting an invasion here.
— Jane Lowrey (Mercer County, clerk)

In the case of Mercer County, Missouri, the citizens were suffering from the delusion that democracy still existed.
— Hugh F. O'Donnell, III (Greenpeace attorney)

We argued that it was wrong to dig up Indians to build a hazardous waste facility.
— Michael Haney (Representative for the Indian Nations of Oklahoma)

One community experiencing a corporate invasion was Mercer County, Missouri. The local people would soon resist the efforts of a huge multinational, Amoco, to site a hazardous waste facility in their county.

Before Amoco Waste-Tech appeared in Mercer County, Missouri, it had a six-year history of attempts to site hazardous waste facilities nationwide, in Ohio, Illinois, Mississippi, Utah, Oklahoma, Texas, Pennsylvania, Florida, and Missouri. In 1990 and 1991, Waste-Tech's efforts were also rejected by several Indian nations, the Chikaloons in Alaska, the Tyoneks in Alaska, the Kaibab-Paiutes in Arizona, the Moapas in Nevada, the Kaws in Oklahoma, and the Navajos in Arizona.[1]

In Missouri, the hazardous waste company had tried three counties prior to Mercer, but in these counties the company withdrew quickly after apparent opposition.[2] This was not true in Mercer County. Waste-Tech persisted in this county for three years, from 1990 to 1993.

Mercer County was typical of such targeted midwestern communities. Small and economically depressed, it bordered on the Iowa state line, midway across the northern tier of counties.[3] A farming area of 454 square miles, with no industry to speak of, the county was devastated by the severe farm crisis of the 1980s. By the time Amoco Waste-Tech became interested in the county as a potential hazardous waste site, more than 20 percent of its population had moved away, leaving just 3,700. As a percentage of the population, this was the greatest demographic shift of any county in the state during the same period.

Business had also fallen on hard times. The wholesale trade in Mercer County in 1987 of just over $15 million represented a plunge of 45 percent since 1982. The county listed only one manufacturing establishment in 1982, coming in last among all counties in the state.[4] With family farming on the decline, corporate farming was also being explored by the local Industrial Development Board (IDB) as a possible way of breathing new life into a moribund economy. A large corporate farming enterprise, Premium Standard Farms (PSF), had been contacted about the possibility of a pig farrowing and feeding operation in the county.[5] Poverty had also risen sharply, reflecting Reagan administration policies toward rural America. The size of the graduating classes in Princeton High School was around 30, down from 60 in the 1960s, reflecting the out migration of the younger population.

In Mercer County, the Waste-Tech agenda was espoused fully by one local leader, Mercer County Industrial Development Board president, Jerrold Taylor. Taylor reportedly told the Princeton *Post-Telegraph* that development of a "complete waste management park"[6] was one of the efforts of the board, along with attracting agricultural firms and other industries. Waste management, according to Taylor, would contribute to solving two national concerns: "One is the environment and the other is debt reduction… We truly believe that this is going to be the industry of the future," Taylor declared, "and we hope to be in on the ground floor."[7]

Citizens for a Clean Environment

If Waste-Tech soon dug its heels in, so did the people of Mercer County. Contrary to the expectations of Waste-Tech, a grassroots movement along with strong local leadership emerged. Protesters soon began employing a number of strategies used by grassroots environmental organizations in stopping unwanted toxic waste facilities—strategies that were effective in building a broad-based opposition that ultimately defeated the project.

Rod Jermanovich, local farmer, attended the regular meeting of the Mercer County Commission on March 12, 1990, and accused the commissioners of meeting with the Industrial Development Board and "trying to operate under the rug" with reference to the Waste-Tech project. The commissioners responded that they were simply "gathering facts and not trying to hide anything from the public." They stated that they had "in mind the best interest of Mercer County and the people who live here and only want to economically boost the county and in no way damage the residents or land."[8]

According to the Waste-Tech Project Proposal, this meant employing 25 to 30 full-time people, with contracting as well as sub-contracting work available to local residents and companies, and the direct economic benefit amounting to some $4 million to the county. If these features did not convince the local citizens, the Princeton *Post-Telegraph* claimed, "Waste-Tech Services' community information programs are designed to answer the questions and bring the proposal into perspective for all the people in Mercer County and other interested communities."[9]

Public relations efforts included double-page spreads in the local paper, summer job advertisements, and flower baskets around the Princeton Square. Early in the struggle, the Waste-Tech public relations specialist for Mercer County was observed on the Princeton Square with a bottle of ash from a toxic waste incinerator informing people that it was "safe" and "edible." Such disinformation angered citizens who opposed the facility for health and environmental reasons.

Jermanovich's response was to organize the Citizens for a Clean Environment (CCE), with the first objective a showdown with the Mercer County Commission. Jermanovich believed in a strong united front, in the importance of mass mobilization. Petition carriers combed the county for signatures over a two-week period, with the petition campaign "Citizens against a hazardous waste facility in Mercer County."

On June 25, 1990, at the weekly Monday morning meeting of the County Court, in a crowded Princeton courtroom, a three-page resolution was read aloud by the vice president of the CCE and 2,200 signatures presented to the commissioners. The presiding commissioner responded, "I feel as Lincoln must have felt at the Gettysburg Address. . . . These people are people who are your friends and my friends. These are people I go to church with and pray with, work with and live with, and I will sign this resolution."[10]

But the First District Commissioner, who would later be known as a Waste-Tech supporter, claimed that the commissioners and community needed to hear both sides before making a decision. The vice president of the CCE replied that Waste-Tech had an unfair advantage: "We've had an unfair battle. A decision needs to be made now." The Second District Commissioner then indicated he would sign, and the First District Commissioner also signed.[11]

A few days later, the hazardous waste company invited some 50 community leaders and elected officials to a luncheon at the Princeton Cupboard to discuss the company's proposal to site the landfill in Mercer County. Approximately a hundred local people took advantage of this promotional meeting—one meant only for the "leaders" of the community—as a ripe opportunity for protest, with placards and chants. They waved "no dump" banners, sang "America the Beautiful," and displayed their 101–page petition opposing the landfill. They did not intend to allow Waste-Tech supporters to turn the county into "Taylor's Toxic Town." Though excoriated in the local newspaper, those who demonstrated against Waste-Tech at the Cupboard defended themselves by pointing out that the action was peaceful. To protest was their constitutional right. The media coverage of this event further polarized the community.

For some time, Mercer County, a generally quiet farming community, would be radically changed by what became an extremely divisive issue. Bumper stickers, hats, and signs brought public attention to the conflict. Bumper stickers read "No Hazardous Waste"; ball caps read "Fight Waste-Tech or Haul Ash" and "Hazardous Waste Is a Slow Death." Signs were pervasive: "Say No to Waste-Tech/Toxic Time Bomb"; "Waste-Tech Want Not"; "Waste-Tech's Toxic Waste Dump Is a Pain in the Ash." Later in the struggle, a billboard on Highway 65, at the entrance to Princeton, read: "Fight Waste-Tech and the Politicians Who Don't." The many signs expressing the people's opposition made "this part of Missouri famous overnight," stated Ed King, New Environmental Winds activist. To the opposition, they were "psychological detriment." The opposition knew how effective signs with political messages can be, and they eventually worked to outlaw them—but, according to King, they were caught in a contradictory position: "If our signs were unlawful, then the politicians' signs were too."[12]

Mercer County soon became host to weekly township meetings, organized and directed by the CCE, with turnouts of two to three hundred people. The CCE mustered support from people from Putnam County, fighting Farmland, a large hog confinement operation; from Schuyler County, fighting Missouri Mining's plans to site a landfill; and from Grundy County, fighting a local landfill bringing in out-of-state garbage from New Jersey. Speakers were brought in on a variety of topics such as the health hazards regarding landfills and incinerators, the threat of property values decreasing in the vicinity of hazardous waste facilities, and strategies for preventing a waste facility from siting. At these area-wide meetings, workers distributed environmental literature and fact sheets. Citizens were encouraged to write letters to local, state, and federal officials and urged to send their opinions to the local paper as well as to the corporate office of Amoco Waste-Tech, in Golden, Colorado. A video camera ran, and buckets were passed for donations, garnering up to several hundred dollars. In Mercer County, one might have expected a turnout of this size for a sports event or a public school function, but not for a politicized event of this nature.

As the struggle unfolded, the *Post-Telegraph* quickly responded with negative editorial comments on any opposition to the Waste-Tech plan. At one point, readers were advised that they were being subjected to "slickly presented one-sided arguments about an issue that . . . was . . . an emotional appeal to our concern about the health of the old and the young alike in our community." The local media's handling of environmental concerns was viewed by many as insulting trivialization of serious issues. By August 1990, just a few months into the struggle, the Princeton *Post-Telegraph* would not print protesters' letters to the editor. The editor stated that "we will not allow ourselves or the newspaper to be manipulated by private individuals or special interest groups."[13]

Other avenues, however, were possible. Individuals expressed their opposition by paying for ad space in other local newspapers. Protesters turned to regional urban newspapers, the *St. Joseph News-Press /Gazette* (Missouri) and the *Kansas City Star*, as well as radio and TV stations, from Kansas City to Des Moines.[14] In addition, like other environmental groups, the people fighting Waste-Tech depended on mass mailings and newsletters, local as well as regional. The CCE obtained a bulk mailing permit and engaged in mailing environmental literature to over two thousand residents in the area. This was soon followed by the CCE newsletter—and by 1991, the newsletter of New Environmental Winds (NEW), a second environmental organization to spring up in the county.

Two special documents also provided a public record of the Mercer County people's position on the Amoco Waste-Tech issue. The first, publication of the names of the signers of the petition presented to the Mercer County Commission, appeared in a paid double-page spread in the *Trenton Republican Times*, on October 1, 1990. The second was a document marked simply *The Letters*, a bulk mailing to all Mercer County residents in July 1992. This document included letters of support for the Waste-Tech project by local residents, taken from the files of the Missouri Department of Natural Resources (DNR). It produced heated controversy. On file down in the DNR's office—as the brief introduction to *The Letters* pointed out—were some dozen letters in favor of the proposal inundated by a stack of letters from the opposition over a foot high. The point was quite clear: These were official documents from the DNR's own files—public information. They were evidence of very weak support for the Waste-Tech cause, with evidence of mass mobilization on the part of the opposition.

Mobilizing Mercer County

From early in the struggle, the people of Mercer County worked steadily to achieve this reality of mass mobilization—employing every possible battle strategy in mind. They conducted their own research on hazardous waste industries and facilities. They made out-of-state trips to see how citizens in areas with waste facilities dealt with their problems. They gathered their own newspaper clippings on daily developments, dating them, filing them away, or passing them on to others. Such documentation of the struggle became a pervasive feature of grassroots activism in Mercer County, as in other grassroots struggles.

The CCE board, as well as its various committees,[15] met regularly to plan township meetings and to develop a long-range timetable. One of President Jermanovich's key goals was to make Waste-Tech such a critical issue that it alone could decide a politician's political future. Since the Waste-Tech project became public in 1990, an election year, the hazardous waste issue could be emphasized as the principal issue in the upcoming election. To prepare for the August primary, the CCE conducted a "Voter's Information Survey." Candidates were asked: "In your opinion is the landfill proposal by Waste-Tech the most important issue in Mercer County?" "Do you believe the Welfare of the People should be the Supreme Law?" "Do you feel the end justifies the means?" "Would you support a hazardous waste facility coming to Mercer County?"

Results of this survey, along with literature informing voters of the dangers presented by waste facilities, were mailed to more than a thousand homes prior to the election. The Mercer County PAC, an offshoot of the CCE, quickly formed and sent out its own campaign literature, urging voters to "Say No" to a slate of pro-Waste-Tech candidates. The results of the election were clearly in the CCE's favor. The turnout for the primary was an unprecedented 51 percent, and two known Waste-Tech supporters were defeated, losing positions they had held for a considerable time. The toxics issue had transformed the election into a one-issue contest.

Pro-Waste-Tech leaders retaliated by charging that the Mercer County PAC's mailing to county voters was illegal because the PAC had failed to register as a PAC as required by the 1985 Missouri Campaign Finance Disclosure Law.[16] Though the Mercer County PAC, politically naive, had unwittingly violated this law, the CCE and the PAC had jointly demonstrated they could use the political process to achieve their goals. In the November election, the Mercer County PAC helped defeat Jerrold Taylor for state representative.[17] The PAC ran a series of 30-second political ads by citizens in favor of the incumbent and opposed to Taylor, whose platform included "revitalization of rural America" and "cleaning up the environment with state-of-the-art technology." The energy of the Mercer and Grundy County resistance in this election was directed to defeating Taylor due to the unsettling prospect of their state representative being a staunch supporter of the Amoco Waste-Tech cause.

At the state level, the Democratic contender, Steve Danner, defeated the Republican incumbent, David Doctorian, who did not come out strongly enough against Waste-Tech. In 1992, the Sixth District congressional representative incumbent, Tom Coleman, was replaced by Pat Danner,[18] who supported those fighting Waste-Tech. Clearly the Waste-Tech issue became a political stumbling block for elected officials perceived to be in the pocket of the toxic waste company.

Besides its political objectives, the CCE had on its planning board a legal mechanism: rural planning and zoning. Historically, north Missouri has not been very receptive to rural planning and zoning since its inception in the 1960s. Mercer County was no exception. The residents were independent, scornful of governmental red tape, and suspicious of interference. Prior to the advent of Waste-Tech, most Mercer countians would have seen no need for such controls on land use, and they would have feared zoning as a threat to agriculture.

The two traditional methods used to site a hazardous waste facility in the U.S. and Canada are "preemption" and "compensation," based on the regulatory or market approaches—or a combination of the two. In the regulatory approach, the state itself, or province, controls the whole process, while in the market-driven approach, the state functions mainly to establish "guidelines for safety and facility operation." Preemption includes, among other things, state statutes prohibiting, or severely limiting, local control through land use laws. Compensation amounts to "bargaining that emphasizes economic enticements for cooperation."[19] In Missouri, the method followed is the market approach, with the public given nominal involvement through a public hearing, once the permitting process conducted by the Department of Natural Resources is relatively complete. What happened in the Waste-Tech siting project was a combination of two methods, with "compensation" being touted as the benefit package by Waste-Tech itself but with de facto preemption the strategic tool used by the state. This preemption by the state became abundantly obvious as the struggle proceeded, as the "fix" clearly was in all the way to the governor's office.

Zoning in relation to hazardous waste facility siting usually comes under the purview of state zoning regulations, federal requirements being minimal.[20] Some states preempt local approval requirements but include "procedural mechanisms" to address local concerns.[21] Most states do provide "some degree of local regulation. In the context of siting controversial land uses such as hazardous waste facilities, courts hesitate to imply the preemption of local land use controls, such as locational or buffer standards."[22] Waste facilities most generally require conditional use permits.[23] On the other hand, "outright prohibition of hazardous waste facilities throughout an entire community, whether expressly stated or a de facto result of the regulations, is susceptible to invalidation on Commerce Clause grounds."[24] Missouri law, under "Environmental Control," makes it unlawful, "with the exception of local option on location" for any political subdivision of the state to "prevent the location or operation of a hazardous waste facility."[25] Missouri hazardous waste law thus does not entirely preempt local control of hazardous waste facilities, but it certainly does heavily restrict local control.[26] According to Missouri law—as well as the Commerce Clause—Washington Township could not "zone out" a waste facility or be so restrictive as to "prevent" a waste facility from locating or operating in the township.

The Washington Township zoning regulations did not explicitly prohibit or "zone out" a hazardous waste facility there, but clearly the regulations were intended to be restrictive enough to "prevent" Waste-Tech from siting its hazardous waste facility in Mercer County. In doing so, the commission was risking invalidation not only by Missouri hazardous waste law but also by the Commerce Clause.

The CCE published a series of newsletters on zoning, attempting to work against "rumors, half-truths, and statements out of context," especially in regard to Missouri law, which specifically prohibits zoning restrictions on agriculture. They identified the process used in creating planning and zoning, emphasized the issue of citizen input into this process, and rebutted the local paper's statement that zoning would increase taxes. They emphasized that the purpose of zoning was to "protect our quality of life" and to "help provide the weapons by which we can defend ourselves from objectionable companies or situations." In one newsletter, the editor wrote that in the past planning and zoning was not necessary "because these were small farms of responsible citizens who would not adversely affect their neighbor. We now have an enormous corporation that wants to dig a giant hole in the ground and dump hazardous waste into it. They don't want a permanent attachment to the land, only 25 years."[27]

The CCE conducted a petition drive and collected enough signatures to place zoning on the November 1990 ballot in the county itself as well as in three out of nine townships. Initially, zoning passed in only one township, Madison, but by April 1991 it had passed in all three, including Washington Township, the location of the Waste-Tech site. Washington Township zoning meetings soon became the focus of the conflict.

Even before the zoning meetings began, in early May 1991, it was apparent that the Washington Township Planning and Zoning

Commission[28] had to deal with a second divisive issue: Premium Standard Farms (PSF), a relatively new factory farming hog operation in the county—and specifically, feeder floors the company had ostensibly planned for Washington Township. To some citizens, PSF was everyday pig farming; to others, it was another invasive force, corporate farming that wished to grow at the expense of local citizens, with little respect for the traditions of the county and the threat of future pollution of streams and groundwater.

At the April Washington Township board meeting, PSF representatives had made reference to a $500,000 surety bond on livestock sewage lagoons in the Madison Township zoning regulations and informed the board that such a surety bond requirement in Washington Township would discourage PSF from moving into the township. This issue dominated the township zoning meetings and hearings for the next two months. Whereas one might have expected vocal resistance to restrictions placed on hazardous waste facilities, instead, in Washington Township the opposition was to a surety bond, of any size, placed on livestock sewage lagoons. Those who voiced this concern were official representatives of PSF as well as those who hoped to engage in consignment farming running feeder floor operations.

The Planning and Zoning Commission viewed such strong support for PSF as a wedge to defeat zoning in the township and thus to clear the way for Amoco Waste-Tech. If zoning could be seen as "against agriculture," then at the August election enough opposition could be mounted to terminate Washington Township planning and zoning. The dilemma for the commission was the choice between PSF and Waste-Tech, both perceived as substantial environmental polluters.

Once the surety bond matter on livestock sewage lagoons, as well as distance requirements from lagoons to residences, was finally resolved,[29] the commission could move on to its principal agenda: developing a zoning ordinance that would, in effect, rule out the presence of Waste-Tech in the township—and since Washington Township was Amoco's one and only planned site, essentially defeating the hazardous waste facility proposal. The method of accomplishing this objective was to take advantage of Section 260.430.2 RSMo, which, as James G. Trimble, attorney for Washington Township, pointed out in a letter to the Department of Natural Resources, "restricts local control over hazardous waste facilities, 'with the exception of local option on locations.'"[30]

The Washington Township Planning and Zoning Commission made use of the "local option on location" provision as its key—and hence most legally defensible—measure of control. Working under this provision, the commission developed two agricultural land use districts, A-1 and A-2, with A-1 being "Agricultural/Recreational/Historical" and A-2 being simply "Agricultural."[31] A-1 was carved out of sections of roughly two miles on either side of the Weldon Fork of the Grand River, an area of land which included the Waste-Tech site. The purpose of A-1 was "To provide a district whose primary purpose not only includes agricultural uses but also preserves and protects recreation opportunities and historical and/or archeological sites."[32] As the CCE July 1992 newsletter stated, "This area is ecologically sensitive, historically significant, especially in light of recent archaeological discoveries [Indian burial sites], and provides a certain amount of recreation to area residents through boating and fishing. . . . This area is a particularly inappropriate area for any type of industry or activity that could cause pollution or that would jeopardize the resources of this area." Waste facilities, hazardous or solid, could be placed in the other agricultural district, A-2, under a conditional use permit, but not in A-1; hence, "local option on location." Clearly, with this zoning decision, Waste-Tech would be prevented from siting a facility in its present location, or in Washington Township at all, since it relied on the river for its operation.

The commission also included, as its second measure of local control—perhaps the more legally questionable one—heavily restrictive provisions in A-2 against waste storage facilities, both hazardous and solid, as well as commercial incinerators handling all kinds of wastes. The Madison Township zoning regulations provided a model for storage facilities, with an extremely creative, and provocative, section on hazardous waste storage borrowed from the Schuyler County ordinance[33] that called for storage in above-ground barrels, with a surety bond of one million dollars per barrel, as well as an annual fee of one hundred thousand dollars from the generator of such hazardous waste. The incineration section was adopted from a very restrictive incineration policy that George Baggett, environmental engineer and activist, was in the process of writing for St. Louis County.[34] Though Waste-Tech continued to deny plans for an incinerator in Mercer County, most protesters took it as an article of faith that an incinerator was definitely in the works.[35]

Amoco Waste-Tech claimed, with support from the DNR, that it did not need to deal with local zoning, period, since its letter of intent to file its application (May 1991) preceded the zoning regulations (July 1991), and thus Waste-Tech was "grandfathered in."[36] Zoners, however, relied on a precedent established in Lafayette County, where zoning was upheld because the waste facility was not actually in operation before the zoning ordinance was adopted.[37] If Waste-Tech had, at some point, been required by the DNR to "comply with local zoning regulations," as Trimble argued they must, the matter would have come down to whether or not Washington Township zoners had taken undue liberties with Section 260.430.2 RSMo, which, as Trimble, indicated, "restricts local control over hazardous waste facilities."[38] But Waste-Tech pulled its application before the zoning question was resolved. For the time being, zoning, in the view of the people, meant local control, a bulwark against the moneyed interests of a giant corporation.

In 1991, under the aegis of New Environmental Winds, a nonbinding referendum became the second legal mechanism for expressing the will of the people. Waste-Tech vice president Doug Johnson was quoted in an article in *Rural Electrification Magazine* as stating, "If everybody here was against it [the Waste-Tech facility], it would probably influence our decision."[39] The people of Mercer County decided to hold the company to its word. If Waste-Tech went forward with the application, it would need to deal with its blatant promise breaking. The nonbinding referendum would, in addition, function to further mobilize those who were possibly on the fence on the Waste-Tech issue. It would be democracy in action, giving the people of Mercer County a voice—even though the referendum would be nonbinding.

On February 4, 1991, the Mercer County commissioners placed the following question on the ballot: "Do you favor a hazardous waste ash landfill and/or incinerator locating in Mercer County, Missouri?" The president of New Environmental Winds, Don Geisse, presented the county clerk, Jane Lowrey, and county commissioners a petition with the signatures of nearly 400 local voters. Lowrey reportedly told the commissioners that she had the authority to place the initiative on the ballot and would do so if the commissioners would not.[40] The county clerk believed that Mercer County would be following a precedent set by St. Louis County, where an incinerator question was put on the ballot the previous November.[41] In fact, the Missouri Constitution specifies that charter counties, such as St. Louis County, are authorized to place initiatives on the ballot, while third-class, non-chartered counties, with less than 5,000 population, such as Mercer County, are not specifically authorized to do so.

While the question of what the courts would decide about the legality of placing the referendum on the ballot remained, the county clerk was willing to test it. The Citizens for a Clean Environment requested that state senator Steve Danner assist them in obtaining an opinion from Missouri attorney general William L. Webster on the legality of the non-binding referendum.

The attorney general's decision on whether a third-class county could place a nonbinding referendum on the ballot was issued on February 21, 1991. The decision concluded, "It is the opinion of this office that Mercer County is not authorized to conduct a nonbinding referendum on whether the voters favor a hazardous waste ash landfill and/or incinerator locating in Mercer County."[42] The decision was based on three precedents.[43] While there was no law specifically authorizing the referendum, the measure could have remained on the ballot. The attorney general's "response was an opinion, and not an explicit order to take the referendum off the ballot."[44]

On March 4, 1991, the commissioners abruptly announced that they wanted the hazardous waste referendum off the ballot. Nonetheless, the county clerk refused to remove it. On March 22, 1991, two of Waste-Tech's strongest supporters filed a petition in Mercer County Court asking the court to order the county clerk to remove the proposition from the ballot.[45]

The hearing was held before a Harrison County associate circuit judge, with the Mercer County clerk represented by Greenpeace attorney Hugh F. O'Donnell, III. O'Donnell argued that "removing the advisory proposition from the ballot was contrary to the principles upon which this country was built . . . that it deprived the voters of Mercer County of their First Amendment rights of freedom of expression."[46] The judge came to a different conclusion. In his decision prohibiting the placement of the referendum on the ballot, he stated, "irreparable harm was shown in that placing the issue on the ballot was use of tax funds. . . . No laws permit such a non binding referendum. I don't know to whom it is even to be advisory."[47]

When the court case removed the referendum from the ballot, the people did not give up but turned to their legislators. Specific legislation prepared by Senator Danner and the local state representative, Phil Tate, provided for a one-time nonbinding referendum on the issue. It appeared quite unlikely, however, that the bills could get through the legislative process and become law before the April 2, 1991, election. In spite of Danner's prediction that it would be very difficult to pass the bill in the state senate, the bill passed in the Senate, 31–1, and in the House, 152–6. New Environmental Winds conducted a letter-writing campaign urging Governor John Ashcroft to sign the bill into law.[48] The bill was sent to Ashcroft on May 13, 1991. He had until May 28 to sign it.

The governor delayed action, finally vetoing the bill on the last day. "Under the law, such a nonbinding preference election, regarding a pending hazardous waste permit would provide no legal basis for either the granting or denial of a permit. . . . The bill . . . falsely creates an impression that the outcome of the election may determine or influence whether such a facility will be located in the county."[49]

Hugh F. O'Donnell, III found it interesting that "the governor failed to acknowledge the fact that Missourians had been permitted to vote on nonbinding issues in the past and seemed to not be too simple-minded to comprehend the significance of their acts."[50] But a member of Governor Ashcroft's staff explained that the nonbinding referendum, in the governor's view, could not serve as a precedent. "Here we have an agency set up with rules, regulations, and statutes to deal with just such a situation. The Boone County referendum was to give the county commissioners guidance about whether to sell a hospital, a matter where they had discretion. In Mercer County, people going to the polls could have no impact, whichever way they voted."[51] From the perspective of Waste-Tech and the governor of Missouri, the people of Mercer County did not have the technical knowledge necessary to make a competent decision, and so it must be left to the Department of Natural Resources.

Don Geisse of New Environmental Winds aptly observed, "The governor has taken an elitist position against citizens who have lived here four, five, or six generations . . . we have been picked as the sacrifice county." But key Waste-Tech supporter Jerrold Taylor said the veto ensured that the decision on a permit would be made by the Department of Natural Resources "and not on the basis of emotion."[52]

Senator Danner was "taken aback . . . . it makes you wonder if the fix isn't in," he stated. Danner said that a lobbyist for Waste-Tech told him that the company did not object to the bill, but "when hundreds of millions of dollars are involved, things can get sticky. That's why I wonder about a veto coming out of the clear blue sky."[53] O'Donnell observed that the governor's veto was "unjustifiable." The governor believed a referendum might "confuse the voters into believing what they were doing was of some import," O'Donnell wrote. "A more plausible explanation for the veto would be that a multi-million-dollar company such as Amoco, the parent company of Waste-Tech, convinced the governor that Senate Bill 420 was not good for business."[54] For Waste-Tech, as well as the state of Missouri, this was strictly an economic issue, a matter of utilitarian moral calculus, with Mercer County the designated sacrifice zone; the welfare of the people would be subordinated to the profit motive. The local people were incensed at the miscarriage of justice.

Grundy County Joins In

Those who opposed Waste-Tech saw the issue mostly in environmental and health terms, but they could also adopt, when politic, strategies that emphasized economic concerns. Partly for this reason, they were able to extend the struggle to neighboring Grundy County, to include activist representatives from the local food processing plant, Trenton Foods. This division of Carnation operated just twenty miles downriver from the Waste-Tech project site. The plant's union, the American Federation of Grain Millers Local #194, was drawn into the struggle out of concern for the Trenton water supply but also for their jobs.[55] The Waste-Tech facility could jeopardize both. Soon, because of the obvious threat to the plant, workers were able to involve Nestlé, the ultimate corporate parent.[56]

As with the CCE in Mercer County, local government was also Union #194's first target, with members meeting with the Trenton City Council, the Grundy County Commission, and the Public Water Supply District No. 1 of Grundy County. In each case, the local governing body officially opposed the siting of the Waste-Tech facility in Mercer County.[57] These resolutions were soon followed by the Mercer County Commission's opposition statement, which declared that it "has become apparent that there is virtually no support [for the landfill] in the community and the surrounding counties."[58] Thus, in both Mercer and Grundy counties, local government responded to evidence of public resistance to the waste proposal. In Mercer County, the site of the proposed facility, the opposition by the county commissioners was initially divided, but one year later, this division no longer existed. Clearly social and political mobilization began to change the posture of elected public officials. The Princeton City Council and the Chamber of Commerce, however, continued to be divided between Waste-Tech supporters and resisters.

Besides confronting local government, Trenton Foods employees used other grassroots mobilization strategies. They educated the public through letters to the editor in local and regional newspapers. These stressed the disastrous effect on the economy of the loss of the local food plant, loss of nearly 600 jobs, the certain decline in property values, and the serious threat to the Trenton water supply. These effects would ultimately extend beyond the Trenton area. A union committee engaged in fact finding. They obtained an aerial photo of the landfill site from the Mercer County Agricultural Stabilization and Conservation (A.S.C.S.) Office, with the Waste-Tech property clearly outlined. They researched the waste to be landfilled, the volume, the processing procedures, the mode of transportation to the site, and Waste-Tech's environmental record. They studied the Trenton water supply, from river to faucet. As a result of their work, the International Grain Millers Union, at its annual 1991 meeting, resolved to assist Union #194 in their opposition to the Waste-Tech project.

Employees organized a letter-writing campaign to elected officials. They made T-shirts, sweatshirts, and caps to make the struggle visible and to encourage community involvement. Bright fluorescent colored posters with slogans like "Amoco Waste-Tech Is Poison" placed in the windows of homes invited questions and concerns. Since Trenton Foods employees lived in several surrounding counties, they were able to involve people from these communities as well.

Trenton Foods employees found the 1992 North Central Missouri Fair a ripe opportunity for protest. On the back of their float, hung a dummy made of a jogging suit stuffed with newspapers, Nerf ball for head, Halloween mask for face, a sign around its neck announcing, "Hang Waste-Tech's Ash!" A second sign on the float itself said, "Waste-Tech Get Your Ash Out of Town." The response from the Trenton community was positive, with the float winning first place in the non-business category, suggesting that what might have appeared "radical" before the potential loss of jobs and threat to the economy was now perceived as a "mainstream" response. Trenton Foods employees also protested Governor Ashcroft's visit to Chillicothe for his "Thank You, Missouri" tour. They picketed directly across the street from the farmers' market site, where the governor appeared. Their protest centered on Governor Ashcroft's veto of the Mercer County nonbinding referendum.

Strong resistance had now become apparent in two counties. But, in addition, Waste-Tech resisters from Mercer as well as Grundy County were part of a broad coalition of regional organizations: The Green Hills Citizens for a Clean Environment, Waste Information Network, the Kansas City Greens Toxics Working Group, the Gateway Alliance of St. Louis, and the Missouri Environmental Action Network (M.E.A.N.). And, as with other environmental groups fighting toxics, organizers against Waste-Tech realized they must fight on a national basis when opposing multinational corporations. Without the assistance of such national environmental organizations, local environmental groups are left to their own expertise and lack the larger framework of issues and strategies that are clear enough to big players like Amoco. In Mercer County, New Environmental Winds, a non-membership group, was created to operate on this national level, to gather information, then disseminate it to the local community. Randy Ferguson, secretary, worked principally with Greenpeace and Citizens' Clearing House for Hazardous Waste (CCHW). Through CCHW, Ferguson obtained information on bad-boy laws, information that later became an effective weapon in the final defeat of Waste-Tech. CCHW also kept NEW informed about Waste-Tech's movement countrywide.[59]

In addition to keeping continual contact with national environmental organizations, Ferguson joined the Amoco /Waste-Tech Pyramid Network, which included contacts in Louisiana, Nebraska, Florida, and Ohio. The purpose of the Pyramid Network was to keep each networker informed of Waste-Tech's actions at other networking sites. Ferguson claims that without this network, her ability to keep abreast of Waste-Tech's "progress" in other states would have been severely restricted. Such national networks, commented Ferguson, keep local groups "apprised of battle strategies."

Sacred Ground

The battle against Waste-Tech in Mercer County was a year old when Native Americans became involved in the struggle because of the siting of the project on sacred burial grounds. Like most Mercer countians, Randi Ferguson realized there were burial mounds in the Waste-Tech area, along the Weldon Fork of the Grand River. This was legendary as well as documented in one history of Mercer County.[60] But the exact location of the Waste-Tech property was not clear until the summer of 1991. With this information finally in hand, Ferguson contacted Native Americans for a Clean Environment (NACE), who put her in touch with Earl Hatley of the National Toxics Campaign and Michael Haney, representative for the United Indian Nations of Oklahoma (UNIO).

At the time, Michael Haney, a Seminole, represented 300,000 people of the 39 tribal entities of Oklahoma as Repatriation Committee Chairman of UNIO. Haney helped write federal legislation PL 101-601, the Native American Graves Protection and Repatriation Act (NAGPRA), protecting Indian burial sites, and has been active nationwide in protecting these sites.[61] As Repatriation Committee Chairman, Haney was officially charged with the responsibility of guarding and preserving the burial heritage of the Iowa Tribe.

Historically, Native Americans were victims of colonization and genocide in the American version of the enclosure movement, the appropriation and exploitation of the vast resources of the North American continent for capitalist accumulation. It was necessary to obliterate the cultural traditions of Native Americans for the sake of capitalist accumulation in America.[62] Not only political but cultural rights and religious freedoms have been threatened by the forces of industrialization and capitalist development. In the case of Waste-Tech in Mercer County, Native Americans and the people of a rural white community came together to protect these rights and freedoms. It is perhaps not surprising that the efforts of a small, backwater community to preserve their cultural heritage and way of life would find resonance among a people historically decimated by the genocide perpetrated by the peopling of the American continent by Europeans. But more than this, the grave sites of their ancestors were being threatened, and the Iowa tribe, who had ancestral rights to these sites by treaty law, were committed to protecting them.

They had the support of the people fighting Waste-Tech, but they also had the force of the 1990 federal law, NAGPRA, behind them.

What the 1990 federal law basically accomplished, in the words of W. Roger Buffalohead, was to acknowledge "that Indian nations and tribes possess political and cultural rights that the larger society must respect and consider in order to reconcile the past and safeguard the present and future."[63] Waste-Tech's plan to construct a facility on sacred burial grounds may be viewed, then, as a violation of such rights, as yet another instance of disrespect for Native American religion and culture—and for native people as "people." Waste-Tech's actions in Mercer County parallel other environmental actions on Indian land by Waste-Tech and other corporate polluters.[64] These environmental actions suggest a racist attitude toward Indians, and they fit into the larger structure of environmental racism, which has been directed toward African Americans, Native Americans, and other people of color.[65]

A new variable was introduced with the advent of Michael Haney. A book published in 1994, *The Book of Elders*, featured Haney as one future elder,[66] and clearly this role has been evident in Haney's various endeavors. Haney sits on a United Nations committee that overlooks international Indian concerns. He was involved in the American Indian Movement's takeover of Alcatraz Island; he participated in the Longest Walk, from Alcatraz Island to Washington D.C., in 1978, to protest the treatment of Native Americans; and he fought with AIM in the 1973 Wounded Knee incident. In 1992, Haney served on a panel with Suzan Shown Harjo and other Native Americans on an Oprah Winfrey program focusing on racism against Native Americans—a typical role for Haney as educator on Native American culture, religion, and current issues. As an environmentalist, Haney helped stop Waste-Tech's siting of an incinerator on Kaw tribe land in north central Oklahoma.

Haney's major goal was to protect the grave sites of his ancestors; however, if he could also prevent Waste-Tech from siting its landfill, "then we're probably killing two birds with one stone."[67] On a walk-through of the land south of the Waste-Tech site, Haney identified mounds but was not certain that these mounds were graves.[68] State archeologist Michael Weichman discovered a 2,000-year-old village site, which Haney confirmed as being on Waste-Tech land. But official documentation on burial grounds and village sites by the Missouri Department of Natural Resources was not available until a change of administration, in 1992.[69]

Haney and the vice president of governmental affairs for Waste-Tech agreed that an archeological impact survey was required under the 1966 Historical Preservation Act as well as the 1990 federal law Haney helped to draft.[70] However, Waste-Tech claimed that the survey would not in any way affect the proposed siting of the hazardous waste facility. In this position, they encountered no opposition from the DNR. Weichman, senior archeologist for the Office of Historical Preservation, a division of the DNR, stated that the area in question could contain "hundreds if not thousands of both burial grounds and occupational sites."[71]

Weichman, however, quickly played down the importance of possible sites in defeating the Waste-Tech project: "Finding sites there will not stop the project. . . . They're blowing this thing way out of proportion." According to Weichman, "No law protects them from a private company's development. If they want to stop it, they should go get some spotted owls and put them out there. Those laws (protecting endangered species) have more teeth."[72] Weichman did concede that should a survey prove the existence of burial mounds on the site, the company would be required to "mitigate (excavate) these sites."[73]

Weichman's statement introduced the conflict between Missouri law and federal law regarding the burial grounds. In a memo to Waste-Tech, Haney countered Weichman's downplaying of legal penalties for disturbing a Native American burial site. Haney pointed out that the 1990 federal law "provides severe penalties, including fines and prison sentences, for desecrating an Indian (unmarked) grave without permission from a lineal descendant of a tribe." In reference to the senior archeologist's suggestion that the company might be required to "mitigate these sites," Haney stated unequivocally that the Native Americans would never permit excavation, nor would they permit "these sites built around."[74]

The legal issue regarding burial sites was the principal agenda item at a meeting in Kansas City between Haney and Victor Roubideaux, cultural liaison and treasurer for the Iowa tribe, and the Unmarked Human Burials Consultation Committee. At the time of this meeting, Missouri state laws made willful disturbance of an unmarked burial site a class C misdemeanor (in 1992, changed to class D felony). Furthermore, as Haney suggested, the state of Missouri "does not recognize unmarked graves over 100 years old"; for Missouri (as for many states), Indian skeletons over one hundred years old were "archeological specimens."[75] Haney maintained that the state of Missouri was not following the 1990 federal law, that Waste-Tech, having federal contracts, "must be governed by the federal guidelines which prohibit tampering with such sites in any way without the approval of the tribes involved."[76] Weichman and others, including the executive director of the Heart of American Indian Center, contested Haney's argument, claiming that the federal law applied only to sites found on Indian or federal land. The executive director "said the commission has done all it can, given the strength of the state statute."[77] Haney and Roubidoux saw the meeting coming to an impasse. As to the Unmarked Human Burials Consultation Committee, Haney stated, "They are mandated federally to protect these mounds and it seems like they are trying to escape that responsibility."[78]

Meanwhile, Haney-Wanatee Park was constructed for the protection of Indian burial sites "after Indian leaders heard rumors about scavenging from burial mounds that are believed to be on the land."[79] The camp was a small park, close to the bridge over Route D, on the edge of Mill Grove, Missouri, with a teepee, two sweat lodges, and a sacred fire that burned continuously.[80] The camp was to be kept running until "we're satisfied that they [the burial mounds] are protected."[81] This camp was an "Indian presence," a symbol of the Iowa tribe's jurisdiction in the area. Three flags were raised: the American flag, the flag of the Mesquakie, and of the Iowa tribe of Oklahoma.

The camp attracted a number of visiting Native Americans representing 30 or more tribes countrywide: Sioux, Winnebago, Mohawk, and

Mandan (Kansas City); Senecan (New York); Cherokee (Oklahoma); Chippewaha (Michigan); Shawnee and Alahama Coushatta (St. Louis); Kickapoo (Kansas); and Mesquakie (Iowa). News of the Indian camp traveled over the Indian grapevine, through ecological magazines and through "Urban Drums" on KKFI free radio in Kansas City. The Indian camp soon became a primary meeting place for local groups resisting Waste-Tech. This setting also presented significant opportunities for press coverage. Michael Haney stated, "All in all, the little guerrilla theater that I think that some of us came up with got a lot of attention of the media and we were able to get on front pages and newscasts from Des Moines to Kansas City on this issue."[82]

Special attention was focused on Columbus Day, 1991. Ferguson of NEW had the political sense to know that the Indian issue was high-profile. "You can't fight an issue without recognition and exposure," stated Ferguson. "We had one hundred percent support from the press because of the issue itself: the systematic genocide of Indian people since the arrival of Columbus. We have a major oil company desecrating sacred grounds. Environmental racism as well as racist attitudes toward Indian communities as a whole."

If Waste-Tech had not pulled its application, the Iowas had the option of bringing suit against the state of Missouri and Waste-Tech on the planned procedure of unearthing burial grounds on a bluff of the hazardous waste disposal site. But this was a last recourse, and Waste-Tech pulled its application before the Native Americans were obliged to battle the hazardous waste company in federal court.[83] Dick Black, AIM member, states that AIM had given an unequivocal promise to protect Indian mounds on the Waste-Tech site. "They would have had a war," Black claims. "These were the guys at Wounded Knee for 63 days."

The Bad-Boy Law

The strategy used against Waste-Tech which arguably caused the hazardous waste company to pull its application was Missouri's Habitual Violators Disclosure Requirement or "bad-boy law."[84] This law has been used in several states, and in Waste-Tech's case, it was the connection between Waste-Tech and its parent, Amoco, and its numerous violations at the Sugar Creek refinery in Kansas City, that clearly placed Waste-Tech in the "habitual violator" category: a corporate undesirable in the state of Missouri, and hence not an acceptable risk for a hazardous waste facility permit.

Bad-boy laws are of three types: debarment, dissolution and revocation, and permit. In 1996, 19 states had the permit type, with various kinds of restrictions.[85] One problem is that bad-boy laws are applied differently toward corporations, depending on their size and influence.[86] What happened in the Waste-Tech case was a product of determined local struggle, and the political clout of a state senator, who insisted that Missouri's bad-boy law be enforced. Use of the bad-boy law was effective enough in the Waste-Tech struggle to warrant a write-up in *What Works 2: Legal Solutions to Toxic Pollution*.[87]

According to Missouri law, "A license or permit shall not be issued to any person who is determined by the department to habitually engage in or to have habitually engaged in hazardous waste management practices which pose a threat to the health of humans or the environment." A habitual violator, according to Missouri's code of state regulations governing owners and operators of hazardous waste facilities, is defined as "a person who has two (2) or more criminal convictions or who has been determined by the department through a two (2)-tier evaluation process to have established a pattern of habitual violations."[88] In October 1991, when Waste-Tech filed its application for a permit, it declared itself on the Habitual Violators Disclosure Statement (HVDS) "wholly owned by Amoco Oil Holding Company" and indicated just two environmental violations by the Waste-Tech Lake Charles, Louisiana, facility—mainly failures to maintain accurate records and to make particular notifications to the Louisiana Department of Environmental Quality. As to the various Amoco enterprises under the holding company, "None of these companies has any ownership interest, nor any management function with regard to WTS." The fact is, however, that Waste-Tech's parent was Amoco

Oil, a corporation with many more violations than Waste-Tech itself had in the state of Louisiana.[89]

On October 16, 1992, an environmental engineer of the DNR Permits Section wrote to Waste-Tech requesting a "modification" of the HVDS to name the "ultimate corporate parent"; it was "MDNR's belief" that this parent was Amoco Corporation, not Amoco Oil Holding Company. Waste-Tech was given 30 days from the receipt of the letter to "submit the appropriate information." Ed King, New Environmental Winds activist, had come to the same conclusion about Waste-Tech's relationship to Amoco Oil. He had also obtained a copy of the MDNR document from his lawyer—just the first page, though, because oddly enough, the second page, which set forth the 30-day deadline, was missing from the DNR files. King appealed the HDVS matter to Senator Danner's office, referring to the HDVS as "an outright attempt to avoid admitting that Amoco has ownership interest in Waste-Tech"—and thus dissociating Waste-Tech from "Amoco violations nationwide," which "would be voluminous to say the least."[90] In a letter to the DNR director, Danner called for an investigation of the questions raised by King and, if "correct," requested "that the Department of Natural Resources suspend the application process for failure of Waste Tech Services, Inc. to make full disclosure."[91] Danner told the press, "Unless DNR can prove his interpretation is wrong, he will take the matter to Governor Mel Carnahan." According to Danner, with Amoco as parent company, 21 Superfund cleanup sites would "make it a habitual violator."[92]

King pointed out to Danner's office that certain letters relating to the disclosure statement issue were missing in the DNR files on King's January 1993 visit, including the one written by the environmental engineer in the permits section, suggesting a DNR "cover-up." This engineer, King claimed, was "on the right track, but was squelched by someone." In King's opinion, "there was an outright attempt by DNR to keep this letter out of public view until after the review was passed back to Waste-Tech," and the DNR had clearly "betrayed the public's trust in this matter."[93]

On Wednesday, February 24, one day before Waste-Tech pulled its application, the press revealed that Waste-Tech had "until Friday to submit the environmental record of its parent company, Amoco Oil."[94] Waste-Tech claimed that it had pulled its application due to "market forces," not citizen opposition.[95] According to then Waste-Tech President J. H. Brinly, due to a slowing down of incinerator permitting in the previous three years, market conditions did not look favorable for a facility "primarily designed to handle ash from incinerators."[96] This "excuse" did not "hold water" for Ed King: "Very likely, non-stop opposition by local citizens had something to do with their decision. It is also reasonable to assume that Waste-Tech did not want to reveal the operating record of parent and sister organizations as required by Missouri law, thereby shielding the dismal record of the Amoco Sugar Creek Refinery cleanup in Kansas City and 19 other Superfund Clean-up Sites."[97] One may well conclude that Waste-Tech was serving as a shell corporation for Amoco, attempting to avoid Missouri's bad-boy law.[98] What ultimately defeated Waste-Tech in Mercer County may be "pure conjecture," as Ed King indicated.[99] But clearly Missouri's bad-boy law, like bad-boy laws in other states, can be an effective tool.[100]

5
Lessons from Mercer County

*If you assume that nothing can be done, then you guarantee
that the worse will happpen.*

—Noam Chomsky

Thomas Gladwin has noted that "there has been a shift in environmental targets from preexisting to proposed industrial facilities."[1]
Gladwin has noted a number of more specific trends: a gradually
expanding role for local government bodies; increasingly greater involvement of local residents; and more frequent grassroots mobilization and
stronger representation by national environmental groups. For instance,
in plans for an underground mine in Wisconsin, the opposition challenged Exxon's prospecting permit, persuaded the county board to pass a
resolution opposing a pipeline, used zoning to pass a socioeconomic mitigation ordinance, and ran a shareholder campaign.[2]

The Mercer County example, fitting Gladwin's second "environmental
target" category (the proposed facility), is chiefly of interest in illustrating
how people mobilized against a major corporation's planned project—and
ultimately won. What their story illustrates is how the inhabitants of a designated "sacrifice zone" ultimately triumphed in the face of a political cli-
mate—at both state and federal levels—that sought to protect and enhance
capitalist profits. In the case of Mercer County, the forces of capitalist accumulation eventually came up short: Amoco Waste-Tech was unable to site
the hazardous waste facility. The protesters were acquainted enough with
the research on incinerators and waste storage facilities to understand that

environmental protections under the regulations of the EPA and the Missouri Department of Natural Resources would do little to make their area safe. Protesters employed the strategies of many other environmental organizations struggling in similar battles since the early 1980s. They mobilized a strong opposition, from many quarters, to resist the moneyed interests of Amoco. They refused to be another victim of the historical enclosure movement; rather, they would do whatever it took to hold on to their land, their natural resources—their traditional rural way of life.

What finally worked to defeat Waste-Tech was the Missouri bad-boy law. But will this strategy continue to work in the future, given the fact that corporate power increases steadily as basic democratic rights are continually eroded? One clear threat is the audit privilege laws already passed in 24 states and pending in several more.[3] What will work tomorrow will depend on the willingness of citizens to confront, at all levels of government, the corporate onslaught against environmental justice for all. Strong mobilization, whatever the strategy employed, will be essential.

A Rural Sacrifice Zone

The people's struggle against the Amoco Waste-Tech hazardous waste facility in Mercer County, Missouri, was part of a "nascent movement against 'rural discrimination.'"[4] The targeting of Mercer and other poor counties in northern Missouri followed the familiar pattern seen in the rest of the Midwest.[5] The dynamics of site selection were explained by Hugh B. Kaufman, an EPA official and grassroots advocate, in a meeting in Mercer County in August 1990. Companies target areas where people are perceived to be the least likely to mount an environmental struggle against a toxic waste disposal facility and defeat the proposal. Border counties are more likely to be targeted since they are in a weaker position in terms of mounting political resistance to the company. Bordering states may vie with each other for the promised benefits. Small, rural counties with a population of less than 25,000 are often targeted, as are politically conservative areas, with a strong belief in the "free market." Populations above middle age, those with only a high school education or less, and those predominantly Republican, are perceived to be more susceptible to corporate propaganda.[6]

A county with limited economic prospects, Mercer County was ripe for exploitation. Company officials can often sell local leaders on a waste project fairly easily with the argument that this particular option may be the only hope for the area, that the local people simply have nothing to lose. Those who object can be made to appear irrational, "academically ignorant," and ready to sabotage the least hope for economic revival. Pressure will be applied to initially confused citizens to leave everything up to the "experts" and engineers. The company and its supporters, outwardly embracing a utilitarian ethic, oppose what they deride as a hopelessly "radical"—and emotionally driven—stance.

Initially, at least, Amoco Waste-Tech had the support not only of the Mercer County Industrial Development Board and the local Chamber of Commerce but also the Mercer County Commission. At the state level, a hazardous waste site meant meeting the terms of the Capacity Assurance Plan, which required a balance of exports and imports of hazardous waste. A disposal facility handling over a hundred thousand tons of hazardous waste annually would certainly put Missouri in the black in terms of this balance. Missouri could demonstrate to the EPA a clear "capacity" to handle hazardous wastes to be generated within its borders over the next 20 years.[7]

At the local level, the Waste-Tech agenda was advanced by exploiting a rural–urban cleavage, which had traditionally existed in the county. Businesses in Princeton were likely to support Waste-Tech, at least initially, while residents of the county generally opposed it. This rural–urban divide was, in part, a clash of economic interests. Businesses tended to see the new "industry" as the best hope for economic revitalization, while those in the county feared the devaluation, and loss, of their land, with the only option being to sell out and relocate. While farming might not be profitable as a business, it was their life, the only way of life they knew. Many also had historical ties to their land, in some cases running back several generations. They saw their community as primarily agricultural, not industrial, and they did not want to become the "ash can" for more industrial areas of the country.[8]

Waste-Tech supporters could always respond to this argument for social justice, as they certainly did, with the familiar corporate argument that "poisoning is the price of civilization."[9] Like others facing toxic struggles, the people of Mercer County were led to believe that they had a duty to the larger society to allow the disposal of hazardous waste in their county. Corporate propagandists have systematically used the term "NIMBY" (Not In My Backyard) to discredit local community groups that fight toxic facilities. Davies Communications, for instance, has worked to neutralize grassroots groups for their many clients, which have included Mobil Oil, Hyatt Hotels, Exxon, American Express, and Pacific Gas & Electric. Companies contact Davies Communications when "a local community is going to shut down" their business. The firm writes fake letters, on behalf of individuals, to public officials, disguised to appear as if they represent a swell of grassroots support against local groups.[10] Corporations are willing to pay PR firms handsome damage control fees to assure profits and an overall pattern of corporate growth and accumulation.

The pejorative term "NIMBY" is used in the literature for case studies of citizen activism against various local unwanted land uses (LULUs). Indeed, there is reason to believe that the literature "is encumbered and perhaps misdirected by an overreliance on the term."[11] The "classic NIMBY" opponents have been defined as those who "try to accumulate evidence of injury not only to them but to society as a whole."[12] Yet the term "NIMBY" as a "catchall term to label the opposition,"[13] who see a diversity of concerns—including that of health—clearly maligns local grassroots activism. If the term is reductive, it is also misleading, suggesting a project (an enterprise of some kind) that an individual basically values but "not in their own backyard," such as a recycling center.[14] But even if the assumption of value (on the part of one or more of the local citizenry) is correct, the term nonetheless fails to come to grips with the issue of social justice, the fair and equitable distribution of goods, as well as bads.

The use of the term NIMBY to describe the Mercer County struggle against Waste-Tech only serves to obfuscate the issues and divert attention from the democratic rights of people to struggle for a clean and healthy community. Social justice precludes the practice of targeting a specific population, denying this group such basic needs. The targeting of Mercer County as a hazardous waste site illustrates well what naturalist Edward Abbey referred to as an "alien invasion."[15] It is but one example of how the resources of local communities, needed for the perpetuation of corporate profits and capitalist accumulation, come under attack by major corporations. People have a recognized right to resist such an "invasion" and to mobilize politically to protect their property. They have a right to struggle for a clean and healthy place to live, free from toxics emitted from a hazardous waste incinerator, polluted groundwater from leaking toxic waste dumps. Corporations were traditionally chartered under the principle that they must continue to serve the people.[16] However, under the emerging, strident New Right interpretations of the sanctity of absolute privilege for private property, particularly the rights of corporations, companies may well be able to use the law and the courts to do just the opposite. In the case of Mercer County, there is much reason to believe that without the people's strong resistance, they would surely have been stripped of their basic democratic rights.

Mounting a Resistance: Mobilizing

As in many other grassroots toxics struggles, the emergence of the CCE and the subsequent struggle were experientially based. Initially, people were stunned that their community had been targeted as a sacrifice zone for the dumping of hazardous waste. They believed, as the ideology of liberal democracy had led them to believe, that justice was just around the corner. They organized, at first, around the single issue of defeating the proposal for the hazardous waste facility. They did so very quickly, with the attitude that when Waste-Tech was defeated, they would disband their environmental organizations and resume living the way they had lived before the advent of this intruder. Meanwhile, they were ripe for mobilization. They met what Sherry Cable and Charles Cable state as an essential requirement for grassroots mobilization: seeing a problem as "immediate and personal—salient."[17]

The supporters of Waste-Tech were quick to brand them as "radicals," led by outside agitators. Predictably, they attempted to discredit the CCE by labeling its first president, Rod Jermanovich, an "outsider." But Jermanovich, a former resident of California, had made north Missouri his home prior to the advent of Waste-Tech. One may grant that he functioned, in a sense, like the "outside organizer" necessary "to spark the initial action" because he did have a degree in political science and plenty of political acumen. But he was, in fact, not an *outside* organizer.[18] Rather, he was a local landowner working to make a living from his own land in the local area. If the local people were part of a "citizens revolt"—and hence "radical" in challenging the local "authority" or leadership—it is best to see them as part of a larger "mass movement" with strong ties to "traditional and particular group identities."

The Left has tended to disparage such sentiments as the " 'backwaters' of parochialism."[19] However, in Mercer County, as in other midwestern rural toxics struggles, a strong sense of family and place often quicken the people to action. The people opposing Waste-Tech grounded their stance in cultural traditions, sometimes falling back on biblical principles and phraseology. They saw an obligation to the rural way of life that transcended pecuniary and market motives. The theory of the economic man fell short. The vision of a "land ethic" and "intergenerational equity" that emerged from their cultural tradition was solidly grounded in the cultural traditions of the rural community they embraced. Those fighting Waste-Tech believed they had a right to be informed about a potential threat to their community and way of life; they were unwilling to depend on "experts" from either industry or government to decide for them. Such local political activism, when not branded "radical," was disparaged as "emotional"—the irrational response of those not ready to listen to the voice of reason and science.

Opponents of Waste-Tech viewed the waste facility issue as a matter of social justice. They had done enough research to realize the dangers of such waste facilities, and the matter at hand was simply that of a politically weak area being "dumped on" by a huge multinational corporation—a David and Goliath struggle. They felt the same "deep sense of betrayal" that has served to mobilize any number of contaminated communities.[20] Under what serves as cultural imperialism, "urban" middle-class values must dominate "rural" values, which are seen as an anachronism, an obstacle to progress. In economically peripheral communities around the globe, local cultures with their own non-market-driven values are often an impediment to capitalist accumulation. If they do not come into the market willingly, then legal and political devices will be employed to bring them into the fold. Major corporations (class struggle from above) put these devices into place, just as international institutions such as the World Trade Organization force peoples on the periphery of the global economy into the global market.

If the people of Mercer County were, in many cases, poor, their poverty had been "ensured" by the political and economic structures in place—and in an area certainly not poor in resources. In fact, it is the functioning of the wider market that tends more and more to deplete such local communities of capital. Land use is critical, as Stephen Horton makes quite clear, since it is through the control of land use patterns that communities can begin to address the preference of capitalism for exchange values over use values. The discard or waste of the use values of a society militates most severely against small communities such as Mercer County, Missouri.[21] Mercer County's pattern of community organizing is quite in keeping with the concept of building community through the people's empowerment, self-determination, and decision making in relation to such critical matters as land use.

Against all odds, a small group of local activists were able to mobilize the support of the people and ultimately defeat the proposal for a toxic waste site. Government is a typical place for grassroots environmental groups to initiate action.[22] One of the first CCE objectives was to confront the Mercer County Commission with the resolution against the Waste-Tech facility, signed by a majority of Mercer County citizens. This resolution was a success, as were similar mobilizations in Grundy County. But environmental issues are not decided at the local level. Efforts to mobilize political power must go beyond local government to the state and federal levels, to congressional representatives. Through a concerted effort to influence the vote, the CCE gained political support at all levels of government. A Waste-Tech choice for state representative could prove disastrous; thus, considerable effort was given to defeating Jerrold Taylor—cutting the connection between Waste-Tech corporate influence and political office.

In the American system of capitalist democracy, political power is largely controlled by corporate dollars, in part through campaign spending and lobbying. While in theory elected officials "represent" the people, for the most part officials represent corporate America, their fundamental constituency. To some extent, the fact that Americans generally do not bother to vote reflects an awareness of how legislation and policy making reflects the needs of major corporations. Public opinion polls have shown that the American people generally believe that government serves the needs of the powerful, not themselves.[23] However, when people are organized and mobilized, particular officials may be targeted and voted out of office when they can be shown to be clearly in the pockets of toxics corporations and other polluting firms. Which politicians took money from dioxin-producing firms? Grassroots environmental organizations can sometimes target such officials, letting them feel the sting of real democracy. Engaging in the electoral process and defeating officials who are doing the bidding of waste companies constitutes an effective method of grassroots political struggle. This strategy "immediately defines the issue as a political problem, rather than a technical or scientific one," thus inviting the people to get a taste of their own power, for a change.[24]

Environmental battles are fought out largely on the ideological front. Use of the media is critical as a "tool for community education"[25]—a commonplace waste companies depend on. The local newspaper is part of the local power structure, and waste companies use the press as much as possible to "get the word out."[26] The people of Mercer County had to mobilize media support in favor of their own position—and provide a corrective to Waste-Tech's propagandistic one, which was visibly dogged in its persistence. Perhaps as a result of so much published protest, the majority of local residents were not persuaded to view the waste facility as a viable economic development or beneficial opportunity for a county with a flagging economy. Meanwhile, such media attention and protest had to achieve a delicate balancing act: protesters had to keep the eye of the public on the Waste-Tech issue yet guard against the perception that Waste-Tech was no longer a threat due to the belief that the problem had been solved through zoning, opposition by governmental bodies, the involvement of Nestlé USA,[27] the active participation of Native Americans—or as a result of so much public opposition in the media. Overall, the use of the media in environmental battles must be an ongoing mobilizing tool for support, education, and protest—a means of organizing the resistance and increasing political power. Of the two, the former may be more easily achieved; the latter is always problematic, given the mechanisms that operate to deny the people their democratic rights.

As in other toxic struggles, in Mercer County political consciousness arose from the experience of seeking justice through the traditional channels. As Celene Krauss argues, those confronting power from below develop a critical consciousness. When the people confronted industry, the state, and political power with their concerns, they found themselves pushing against a bulldozer headed their way. This was the point at which the reality began to sink in, the understanding of the link between public power and private encroachment. A critical political consciousness emerged in which they began to understand the power nexus between the state and the waste industry and that they and their community and way of life had been "invaded," targeted for capitalist plundering.

The emergence of a political struggle on the local level was, as in other cases, grounded in a popular culture. Populist distrust of big business is indigenous to the Midwest. Historically, there has been a sense among most farmers of having an obligation not to sell out their rural way of life just for short-term monetary gain; this conviction seems to derive from religious teaching and influence. Exploitation at the hands of urban-based corporations was nothing new to the people of north Missouri. The concept was in their basic worldview; after all, much of the capitalist accumulation to build the industrial infrastructure of America was produced by farmers in the fertile and productive lands of the Midwest. Now, not only was their welfare at stake, but this time, their very way of life had been targeted for extinction. The conviction emerged that deeply held traditions and values were being devalued. They were about to become victims of enclosure. They themselves were waste to the invading waste corporation. Political activism, while succeeding in gaining favorable political support (and ousting opponents), failed critically, at times, to empower the people—especially when Waste-Tech appeared to be a fait accompli, facilitating the "unmasking" of the ideology of liberal democracy.

Township Zoning

Zoning, regardless of its limitations, is one strategy citizens may employ in struggling to gain democracy and environmental justice. Land-use zoning is, in a way, a counter-enclosure movement, whereby local communities take back some of the control over the resources that have been stripped away from them. It goes against the thrust of the market and commodification, as well as against environmental degradation. It is contradictory to the inner logic of the market and capitalism, which requires the commodification and marketing of every resource. Land-use zoning is an assertion by local communities that some things are not for sale. In Mercer County, zoning was perceived as a measure of local control, and the commission was determined to create an environmental model that gave them the greatest control possible.[28]

If the matter went into litigation, the court would have had to decide on the township's particular establishment of land use districts and whether an arrangement that would result in a de facto prohibition of the Waste-Tech siting was, in fact, permissible—or simply patent evidence of "neighborhood opposition"[29] to an unwanted intruder. One must consider a rural township pitted against corporate legal challenges in court—with all the financial, legal, and political clout a large corporation has at its disposal to gain the resources it desires. The Waste-Tech struggle never reached this point. Yet if zoning was never actually tested in court, it was nonetheless a strategic device for possible litigation. In terms of social justice, it was the people's voice expressing their will, their belief in self-determination and other key democratic principles outlined in this society's major political and legal documents.

The Nonbinding Referendum

People fighting environmental battles must use all resources available, including legal ones. A nonbinding referendum would give the people of Mercer County the right to cast their ballot on the Waste-Tech issue— simply a declaration of their political position, no more, no less. It would be an effective measure in creating a public document for the power structure in Jefferson City.

Many newspapers have been effectively under the control of corporations and political machines. The *Los Angeles Times* bitterly opposed the movement for democratization fearing that "business and property rights" would be in danger and that "the ignorance and caprice and irresponsibility of the multitudes" would be substituted for "the learning and judgment of the legislature."[30] The *San Francisco Chronicle* also opposed popular democracy, observing that the movement for initiative and referendum went against the best traditions of American government, which had been to limit the "unbridled license of temporary majorities." The editors called upon established elites to "put down the proletarian movements."[31]

State courts have traditionally been hostile to direct legislation,[32] under "the general principle that a body acting under delegated authority cannot redelegate its powers to some other person or body." The vote of the people has often been viewed as a "foreign or extraneous power," "a delegation of power to a foreign power which is not known to the Constitution."[33] Courts have also argued that since the U.S. Constitution provides that Congress should guarantee each state "a republican form of government," any state is prohibited from establishing a "democracy."[34] Courts have viewed the referendum as constitutional only if the written constitutions of states expressly confer such a right upon the people.[35]

The constitutionality of direct legislation in the form of the initiative and referendum was challenged in *Pacific Telephone and Telegraph Co. v. Oregon* in 1912 on the grounds that such direct democracy was a denial of a representative (or republican) form of government. The U.S. Supreme Court declared these forms of direct democracy constitutional in this case. The Court said that letting the people vote on a measure did not prevent the legislature from functioning as a representative body.[36]

Following the populist era struggle, many states adopted provisions for a combination of the direct democracy instruments of the initiative, referendum, and recall.[37] Typically about 60 percent of measures in states win approval.[38] Environmental issues, including restrictions on nuclear power, nuclear waste sites, and toxic waste, have appeared as referendums on ballots. In 1986, 83 percent of voters in Washington state urged Congress to disapprove of a nuclear waste site.[39] In California, the initiative has been used to tighten restrictions on toxic waste. The referendum can sometimes be an effective tool for environmentalists.

While public support for direct democracy was strong in the 1980s,[40] there was an increasing tendency for courts to remove measures from ballots, especially advisory initiatives. State courts often remove such measures from ballots on the grounds that "this was not the intent of the state's initiative and referendum process."[41]

It is ironic that the success of the democratic movement to establish such direct democracy has not resulted in placing greater power in the hands of the people. In practice, the referendum has most often been captured by conservative interests to obstruct what such elites viewed as "radical legislation" by legislatures.[42] Often money and propaganda by large corporations come to be the determining factor, such as in sinking bottle recycling bills, with the referendum used as a tool of those who wish to maintain the status quo.[43] The referendum is a potential tool that can be used in the public interest only when the people are given the right to use it and have the positive resources to make their voices heard above moneyed interests.

The defeat of the nonbinding referendum in the Waste-Tech struggle illustrated the regulatory process by which the pattern of decision making increasingly takes power out of the hands of the people and places it in the hands of "neutral" technicians and engineers. O'Donnell reflected: "I had come to unwittingly accept the erosion of the democratic process in these circumstances. . . . I failed to recognize the injustices in the system. . . . I had become indoctrinated in the regulatory process." Commenting on the governor's veto, he wrote, "In his veto, Governor Ashcroft embarked on a voyage of rhetoric that is typical of what we've come to expect from today's politicians. In an attempt to convince Missourians that Senate Bill 420 was unnecessary, the governor cited the 'broadest public participation requirements of any environmental law.'" In other words, the governor tried to appease the citizenry into accepting the proposition that the regulatory agency, the Department of Natural Resources, would look out for the public's interest. But as O'Donnell pointed out: "Herein lies the fallacy of our present-day regulatory process. The system gives the illusion of democracy in action when such is not the case. . . . [A] public outcry of great force and magnitude . . . will not serve to prevent its approval. The department . . . is required to issue a permit if all the technical requirements . . . are met."[44]

In grassroots environmental struggles, large corporations are often confronted with people who insist upon using every political and legal tool at their disposal, rather than conforming to their expected role of deferring to the economically and politically powerful. Grassroots political struggle through referendums, the use of the courts, local zoning, and bad-boy laws predictably meets stiff opposition, not only from big corporations used to buying whatever they need to continue capitalist accumulation, but from the government also. A useful litmus test for democracy is whether or not institutions exist and function in such a way as to allow the people to decide local public policies.

Legal strategies employed in Mercer County against Waste-Tech were attempts, like other strategies, to gain environmental justice. When the nonbinding referendum was vetoed, neither Danner nor Tate gave up, but introduced yet another nonbinding referendum, and even a binding referendum.[45] If the fate of the nonbinding referendum gave the citizens of Mercer County a real sense of political powerlessness, it was also verification that "the fix was in"—that Mercer County was indeed slated by the governor himself to be a "sacrifice zone." This knowledge could have been quite destructive, but the people continued to mount a resistance, looking to other strategies.

Waste-Tech and the Issue of Environmental Racism

The state of Missouri's position—that the federal law in question (NAGPRA) applied only to sites found on Indian or federal land—suggest at least two possible issues, according to Dick Black, official representative of the Iowa tribe on repatriation in Missouri.[46] The state of Missouri may have wished to maintain that the Iowa tribe had no jurisdiction in the area; the Iowa tribe had yet to prove, at least to the state, that they had aboriginal title in line with NAGPRA, Section 3, regarding ownership of human remains and objects.[47] This should not have been an issue, however, as Black points out; Michael Haney brought to the meeting of the Unmarked Human Burials Consultation Committee documents from the Department of the Interior providing evidence of the Iowa's aboriginal title. Even so, the committee appeared to operate under the assumption that they, not the Iowa tribe, were authorized by Missouri law to handle any matter relating to mounds discovered on Waste-Tech land. It was their position that they need not contact a tribal nation.

Second, the state of Missouri, firmly in support of Waste-Tech, may have wished, Black speculated, to feign ignorance about Indian burial mounds on the site (withholding documentation). Then, when the mounds were disturbed, the state would turn over the matter of "mitigation" (excavation and reburial) to the committee. But to Michael Haney and the Iowa tribe, mitigation was unthinkable and also a violation of federal law. "We attacked them on a moral basis," Haney stated. "We argued that it was wrong to dig up Indians to build a hazardous waste facility."[48]

The treatment of Native Americans as "specimens" instead of "people" has clearly been religious discrimination.[49] For native people, what is at stake here—and what Amoco Waste-Tech chose to ignore—is the Native American concept of the "sanctity of the dead."[50] In clarifying the importance of the 1990 federal law, Tessie Naranjo, chairperson, Native American Graves Protection and Repatriation Act, identifies the fundamental relationship between objects, the dead, and the living in Native American thinking. Human relationships, according to Naranjo, break down into three types: relationships to nature, to the whole natural environment; relationships to others, to ancestors, the present community, and to future native communities; and relationships to things created by humans. The integral relationship between persons and things is central to an understanding of Native American religious thinking regarding skeletal remains and sacred objects:

> Respect for all life elements, including rocks, trees, clay, and pots, is necessary because we understand our inherent and inseparable relationship with every part of our world.
>
> We also honor those people who have gone on and the objects that they created and with which they established an intimate relationship. This perpetual honoring of our ancestors allows us to remember our past and the natural process of transformation—breathing, living, dying, and becoming one with the natural world. We are never unrelated—not even in death.

Narantho points out the curious treatment by nontribal peoples of objects and human remains as "nonliving entities"—as subjects for "study and analysis"—and contrasts this treatment with one that bestows honor "for the human relationships they represent."[51]

One must add to this the Native American belief that the dead are on a journey, and this journey must not be broken. To disturb the bones of Indian dead is to interrupt a spiritual odyssey. To hold collections of bones in museum vaults is to hold ancestors "in captivity."[52]

It was the violation of sacred bonds, as described by Narantho, and the issue of disturbance of the journey of the dead, that were central for Native

Americans during the Waste-Tech event. But a second issue is raised by Narantho: one that concerns a prominent historical pattern of cultural imperialism and commodification by the dominant white culture, of treating native ancestral remains as "nonliving entities." The act of "mitigation" calls up a whole history of Indian bones being used as "specimens" for scientific study.[53] To Native Americans the desecration of grave sites by archaeologists and artifact hunters and the vast holdings of skeletal remains and grave goods by museums have all been part of a history of genocide and racism, a human rights issue.[54] The treatment of Native American burial sites and skeletal remains amounts to "desecration" by a highly secularized culture.[55] In a speech delivered on Columbus Day, 1991, Michael Haney called attention to "that colonial mentality that American Indian people are less than human" and expressed dismay that archeologists funded by both state and federal dollars can "dig up our remains and put us on display" and "dig up our villages for the sanctity of educational opportunity." They can "mine my people like coal and uranium."[56]

Whether intentionally or not, Waste-Tech related to skeletal remains in the context of this long history of viewing Indian bones as "less than human"—as "cultural property." The concept of "property in the public domain" has worked not only in the acquisition of native lands but also in that of intellectual and cultural property.[57] On federal public lands, "cultural property" has historically belonged to the U.S. government. Native American remains at least one hundred years of age were, until the passage of NAGPRA, "archeological resources" under the control of the Secretary of the Interior, protected by the Antiquities Act of 1906 and the Archeological Resources Protection Act of 1979.[58]

The anthropological/archeological perspective on skeletal remains and funerary objects certainly bears mentioning here. In general, this perspective is that without the study of skeletal remains and grave goods, much vital information about Native American cultures would be lost, not only to scholars but also to Native Americans as well.[59] Those who support the holdings of Native American skeletal remains in museums see museums as a "bridge between Indian and contemporary American culture."[60] Paul

G. Bahn and R. W. K. Paterson argued that scientific investigation is extremely important in terms of gaining "worthwhile knowledge" of peoples and their cultures; destruction or reburial without easy access to remains may mean a tragic loss of such information.[61] "The archeologist's dilemma," claimed Bahn and Paterson, "is to reconcile the duty to respect the implicit promise undertaken by a group of living human beings (to preserve the remains of their ancestors from molestation) with the duty to advance our knowledge of past societies and their practices."[62]

Waste-Tech's role in Mercer County was placed, by Michael Haney, among others, in relation to the viewpoint that regards Native American skeletal remains as "archaeological resources." Whether one views archeologists as nobly pursuing a mission of gaining "worthwhile knowledge" of the past or as fitting into a larger structure of human rights violations against Native Americans depends largely on whether one has a "white concept" of burials or a Native American one.[63] But the point is that the Waste-Tech project must be viewed in terms of a larger framework of racial and religious discrimination. For Native Americans in Mercer County, Waste-Tech was allied with archeologists and artifact hunters.

Habitual Violators Disclosure Requirement

While corporations routinely cite market forces as the reason for pulling out of a desired location, this argument is disputed by experienced activists. "The incinerator industry has often claimed that unfavorable 'market forces' have led them to conclude that building an incinerator is a bad idea. In all cases, they would have built their bad idea if not for the efforts of grassroots activists who have organized to stop the protests."[64] At the same time market forces were cited as the reason for canceling the Missouri facility, the Kimball, Nebraska, facility was brought on line. The company operating the facility, Clean Harbors of Braintree, Inc., claimed, in 1999, to have not yet made a profit, but expected to be in the black in the future.[65] The people of Mercer County deserve credit for mounting an effective campaign, obtaining press coverage, and defeating the efforts of a major corporation, against all odds. The final stroke may well have been the use of Missouri's Habitual Violators Disclosure Requirement, or bad-boy law, Missouri's most effective legal strategy against Waste-Tech.

While bad-boy laws (or good-character laws) are on the books in many states, what happened in Mercer County illustrates that only when citizens push for enforcement of these laws are they really effective. Too often large corporations, like Amoco, are given a wink and a nod by state and federal regulatory agencies. Citizens fighting for environmental justice come to understand that governments most often serve the interests of capitalist profits and accumulation as part of the necessary logic of the system. The people must therefore be proactive. They must be "in the face" of the corporate polluter as well as the state regulatory agency—even when provisions like the bad-boy law exist.

However, with nearly half of the states having passed corporate secrecy laws, or audit privilege laws, if the trend continues, grassroots activists may not be able to rely on bad-boy laws in future environmental struggles. Some recent examples suggest that in states where such laws exist, companies may hide their environmentally poor records from citizens, and state regulators, in an attempt to produce favorable results in a permitting process. For instance, Waste Management, even when the audit bill was pending in Ohio, claimed privileged status for its internal audits as it tried, in the meantime, to expand its existing landfill. Waste Management was unsuccessful, but this was perhaps only because the audit privilege law had not yet passed in Ohio.[66] A lack of public access to "permit related information" recently led the EPA to notify the Kentucky Natural Resources and Environmental Protection Cabinet that its Title V Operating Permit Program is deficient. Kentucky has passed both audit privilege and immunity statutes. The EPA's position is "that Kentucky's law impairs the state's ability to adequately administer and enforce its Title V operating permit program as required by the Clean Air Act."[67] These statutes, audit privilege as well as immunity, vary from state to state, yet what remains consistent across the board is the blocking of vital environmental and health information from the public.[68] If citizens cannot obtain permit information, bad-boy laws will surely be jeopardized.[69] The use of the bad-boy law may be left entirely up to the discretion of engineers in state regulatory departments.

Audit privilege laws appear to be a very significant development in terms of concentration and accumulation of capital in the waste industry. On the one hand, the biggest industries support higher standards for waste handling and shut out the smaller companies that cannot meet the capital requirements. On the other hand, the same giant companies escape punishment when the laws are not met and when local groups lack knowledge of violations of environmental laws. The risk is that audit privilege laws may function, even beyond their implicit "pollution secrecy," as "dumping grounds," or cover-ups, for serious violations, even willful, planned ones.[70] For environmentalists, worse yet, some states' immunity statutes could mean that serious violations will go unpunished. Anti-whistle-blower laws further enhance the power of corporations to pollute freely, unless somehow restrained by state regulators or by the EPA. With greenwashing labels such as the "Environmental Protection Partnership Act" in the U.S. Senate and the "Voluntary Environment Self-Evaluation Act" in the House of Representatives, audit privilege bills are supported by huge corporations who say they will improve environmental protection and public health. They simply need to be able to conduct their internal audits without fear of reprisal.[71] Some of the corporations supporting such legislation include AT&T, Caterpillar, Du Pont, Coors Brewing, Eli Lilly, 3M, Pfizer, Procter and Gamble, Weyerhauser, and WMI (the world's largest waste company). With 25 percent of states not enforcing existing environmental laws, it seems likely that audit privilege laws will only further weaken the EPA and state enforcement of the environmental legislation that does exist.[72]

To fight such secrecy laws, the "Network Against Corporate Secrecy" was organized by activists in Cambridge, Massachusetts, as part of the Good Neighbor Project for Sustainable Industries. The network links some 80 organizations nationwide.[73] Those who would fight for the last vestiges of local democracy have their work cut out for them. With the strident ideology of neoliberalism abroad and worming its way into the structure of court decisions, it is not at all certain how long bad-boy laws will survive as the corporate onslaught advances. In the case of Mercer County, the bad-boy law option was available, and still is; however, a number of attempts have been made in recent years to pass an environmental audit privilege law in Missouri.[74] Those who would look to future environmental struggles must keep a close watch on audit privilege and immunity laws in states where these are pending. Defending his own audit privilege bill, Missouri Representative Gary Marble stated: "Because businesses and industries pay more taxes than common citizens . . . they should be granted more rights."[75] This political philosophy is a menace to ordinary citizens.

6
Wasting the World:
Enclosure, Accumulation, and Local
Environmental Struggles on a Global Scale

I think the economic logic behind dumping a load of toxic waste in the lowest-wage country is impeccable and we should face up to that.
— Lawrence Summers

The name toxic imperialism has been used for struggles against the export of toxic waste. Such struggles could easily link up with the Environmental Justice Movement in the United States.[1]
— Juan Martinez-Alier

Looking at the invasion of a local community in the Midwest by a major global corporation leads one back to the crucial question of class. The environmental justice movement has done much to expose the racist and imperialist nature of corporate policies in the United States and around the globe. It has set out an important grassroots-based agenda to address these issues. But in focusing almost exclusively upon ethnic minorities, women, and the Global South, while neglecting small-town white America, it has sometimes failed to get to the root of the problem—the fundamental cleavage dividing American society and societies around the globe, that of class.

The case of Mercer County, with the area's own well-established local cultural traditions, serves to illustrate how capitalism is an "equal opportunity" exploiter of people and communities. Grassroots community struggle is a struggle against the obliteration of a way of life. The Amoco Waste-Tech assault in Missouri was an orchestrated assault upon a small community, an invasion, in which we see the separation of the town and the country and the clash of interests between them, as noted by Marx.

Capitalism has been in the business of "sweeping the agricultural population off the land" for a long time, going back to the historical enclosure movement. In stripping people of their land and livelihoods, it ensures poverty as well as inequality. If the corporate agenda did not succeed in Mercer County, this may be attributed to the strong mobilization of the local community—but perhaps as well, to time and place. With corporate secrecy laws in nearly half the states in the U.S., environmental battles will be fought, at least in these states, on much harder ground.

And such battles must be fought. For the fact remains that the corporate agenda cannot be furthered without the continuation of the historical rape of the earth. It cannot go forward without commodification, degradation, and the destruction of use values in the form of a cornucopia of waste. This wasteland of plenty means massive profits to the ever fewer monopoly waste corporations. To the majority of humankind it does not mean the good life, but the disposal of the good life, the disposal of their land, their cultural traditions, the disposal of the use values they have created, the disposal of the surplus population of the workers and ultimately the disposal of people themselves. It means the "sanitary" disposal of living communities, historical nations and a humane world, in the landfill of history. This is the brave new world of "markets" and neoliberalism. "This is the new Waste Management . . . for the new millennium."[2]

Capitalist accumulation, commodification, and environmental degradation are global in scale. As we have seen in Chapter 2, capitalist accumulation encourages the creation of waste in the preference for exchange values over use values. The use value of local communities and of living cultural traditions is giving way to the capitalist corporate assault all across the globe. As Stephen Horton points out, "local communities are the biggest losers when use-value has to give way to exchange value."[3] This ongoing historical process that we have traced to the Enclosure Movement, the "creative destruction" of capitalism, has been characterized as "taking a meat-axe to living communities."[4] Such cultural imperialism, whether against small white rural communities in the U.S., Native Americans, or tribal peoples in India, provokes local responses and struggles.

A striking example of the global corporate-state agenda may be found in a 1991 World Bank report from Lawrence Summers, then chief economist at the World Bank. Here, Summers, who later became U.S. Treasury Secretary, argued that it is perfectly logical to dump hazardous waste in the lowest wage country. The report saw Africa as "underpolluted."[5] This view of economic rationality, which is a form of toxic imperialism by transnational corporations, is being sharply challenged by people in many Third World countries.[6]

The World Trade Organization and the Global Market in Toxic Waste: The Commerce Clause on a Global Scale

As we have already seen, the Commerce Clause is one of the most powerful instruments of U.S. corporations in consolidating the monopoly waste industry and in increasing capitalist accumulation. In allowing giant, integrated economies of scale, and the monopoly of a scarce commodity, landfill space, waste corporations are becoming more and more powerful. The corporate agenda of Waste Management Incorporated, Allied Industries, and other waste corporations is to extend the unimpeded trade in waste, hazardous or not, on a global scale. The Commerce Clause serves to strip states and local communities of local control and democratic rights to control their communities and their future. The World Trade Organization functions as a Commerce Clause on a global scale, allowing waste companies to move waste at will anywhere in the world. This is what global waste corporations need in order to maximize profits on a global scale. The U.S. is one of six nations, including the U.K., Germany, Australia, Japan, and Canada, that have played a blocking role in the attempt to establish conventions that would restrict the international trade in toxic waste. Just as solid and hazardous waste in the U.S. is considered a commodity in interstate commerce, by the same token, the U.S. supports the treatment of toxic waste as merely a commodity to be traded globally. When corporations are global in their operation, the complete freedom to move waste can contribute significantly to global profits and capitalist accumulation. While the U.S. has the "right to block the importation of goods that fail to comply with its health or environmental standards,"[7] in March 1996, EPA administrator Carol Browner approved the import of toxic PCB wastes from Mexico and Canada for incineration and disposal in waste fills in the U.S.[8]

Under the "Final Rule," the Toxic Substances Control Act would be amended to allow the import of PCB waste for disposal as long as certain conditions were met. According to the EPA, "an economic benefit in the range of $50–100 million annually for the U.S. disposal industry will result." Another benefit could be to "stabilize disposal prices for U.S. PCB waste generators in the future, by ensuring the U.S. PCB disposal facilities continue to have an economically viable market, and continue to remain in the PCB waste disposal business." In asserting that these imports would not "increase the risk of injury to human health or the environment," the EPA pointed out that the 173,000 metric tons of PCB materials in Canada and 60,000 tons in Mexico could be compared to the "842,000 tons of domestic PCB waste" disposed of at U.S. commercial facilities in 1993.[9] The decision to amend the TSCA was later blocked by a court order, but such a decision illustrates U.S. policy on the international trade in hazardous waste.

The international trade in toxic waste is closely linked to foreign debt and a developing country's need for foreign exchange. Hazardous waste has been shipped to developing countries in many forms. It has been shipped under the guise of "humanitarian assistance" in the cases of radioactive milk and illegal pesticides. It has been shipped under the label of fertilizer, landfill material, fuel, recyclables, dry cleaning solvents, road construction materials, battery recycling, and an almost endless list of other ways. In one example, Daimler Benz contracted to ship 50,000 tons of industrial waste sludge from Germany to Turkey as "fuel" for a cement plant in the town of Isparta. The material was falsely labeled as "fuel." Some 1,600 tons of the sludge was shipped in 1987, containing 168 various substances, including lead, chrome, copper, cyanide, dichloromethane, dichloroethylene, and PCBs. Further shipments were halted after analysis of the "fuel" by Middle East Technical University in Ankara identified the material as hazardous waste. The literature abounds with examples of companies using such devices to deceptively dispose of their wastes in the Third World, when disposal at home would be many times more costly. Nongovernmental organizations (NGOs) such as Greenpeace have played a major role in working to ban the trade.[10]

The Basel Convention, which was brought into force in 1992, does not provide adequate protection for nations from the trade in toxic wastes, and some environmental NGOs have called it "legalization of toxic terror." Some 95 countries are a party to the treaty. It was not ratified by the largest waste-producing states nor the largest importing states among developing countries simply because its restrictions, however weak, were perceived as a threat. The largest waste producing states in the West, the U.K., Germany, Australia, Japan, Canada, and the U.S., played a blocking role, as these states saw restrictions in the toxic trade as militating against their national interests. In fact, the United States has led the veto coalition against such treaties. As for the developing countries, African states wanted a total ban on waste exports from developed countries along with export state liability for illegal trading. But they lacked the political clout to get such an agreement at Basel. Led by the United States, the Western countries forced the developing countries at the final meeting in Basel, in March 1989, either to accept an "informed consent regime" or to get nothing. Under the informed consent regime, waste exporters would be required to "notify their governments of any exports and notify importing countries of any shipments before they arrive." Consequently, the treaty did little to limit the trade in toxic wastes. The treaty allowed the developed states to continue the trade, while denying any significant protection for developing states. Among its weaknesses, the treaty seeks to regulate waste rather than ban it; has no provisions for actually stopping shipments; contains vague definitions of the terms "hazardous waste" and "environmentally sound"; and excludes radioactive waste.[11]

From the perspective of the waste and recycling industries, the Basel Convention "damages recycling," and the industry complains that no one can be very sure about just what materials are banned from international trade. This is an argument made by the Bureau of International Recycling (BIR). This organization legally challenges the Basel Convention definitions, saying it hampers the international trade in plastic, paper, textiles, and so on. With recycling operating under low profit margins, the waste companies are likely to be able to mount increasing political challenges to treat hazardous waste as any other commodity in global trade, as the case is with hazardous waste in the United States. The monitoring and control of cross-border transportation of hazardous waste and recyclables are problematic, not only because parties claim they do not know what is banned[12] but because the various definitions and classifications of hazardous waste are themselves political, as we saw in Chapter 1. Under globalization, these dynamics of capitalist logic operate universally.

The 1991 Lome IV Convention between the European Community (EC) and 69 African, Caribbean, and Pacific (ACP) states is a stronger commitment to control the international traffic of toxic wastes. The EC states agreed not to export wastes to the ACP states, and the ACP countries agreed not to import such wastes from any non-EC states. The African Bamako Convention, negotiated in 1991, attempts to ban the import of toxic waste, including radioactive waste, into Africa. Regional bans have also been adopted in some other areas.[13]

These two conventions are worthy efforts in controlling the flow of toxic wastes. But with the U.S. taking the lead in establishing global systems for the tracking of toxic chemicals, similar to the cradle-to-grave system in place under the TSCA in the United States, environmentalists and environmental NGOs clearly have a difficult road ahead of them. The chemical industry and other powerful corporate sectors have been able to protect their most profitable emissions under the TRI system of reporting in the United States. Now several countries are in the process of developing TRI-like systems. These go under the name of "Pollutant Release and Transfer Registers" (PRTRs). Supporting these efforts is the Commission on Environmental Cooperation, which was created by the environmental side agreements of the North American Free Trade Agreement (NAFTA). In 1996, the Organization for Economic Cooperation and Development (OECD) published a guidance document for governments. At least eight countries have set up PRTR systems: Australia, Canada, France, Mexico, Netherlands, Norway, the U.K., and the U.S.[14]

For the Global South, the United Nations Institute for Training and Research (UNITAR) has set out procedures on how to implement a toxics reporting and tracking system. A pilot program has been set up to include Mexico, the Czech Republic, and Egypt under the PRTR Coordination Group, chaired by the U.S. and in cooperation with the OECD, which serves as the secretariat.[15] Such a system may have the potential to increase knowledge about toxic emissions globally. Politicized, it could serve as a vehicle by which the most polluting global corporations might prevent radical and more effective regulations, thereby protecting their most egregious and profitable emissions wherever they operate globally. Some countries in the Global South have higher standards than those in the U.S.—at least, in particular areas. "In some cases, Mexican standards are higher than environmental regulations in the United States."[16] On the other hand, the U.S. recognizes that some regulation is an advantage to U.S. corporations. U.S. Representative Richard Gephardt has argued that "the failure of other states to enforce their environmental laws places foreign manufacturers at a competitive advantage over American producers. ...I refuse to accept the notion that we should lower our standards environmentally so that our companies can compete more effectively."[17] The statement itself suggests the utopian concept of effective "environmental protection" under global competition. Granted, U.S. corporations have polluted in order to compete globally, as is evident in the post–World War II U.S. economy. But it is clear that even moderately higher standards in many countries of the Global South, as well as in former communist states, could contribute to the ability of U.S. multinationals to meet the higher capital requirements of new regulations and thus increase their monopoly power. We have seen how higher standards for waste disposal in the United States have contributed to the power of monopoly waste corporations.

Challenging the Corporate Assault on the Global South

There is an ongoing scramble for global resources, with resulting commodification and environmental degradation—whose roots we have traced to the historical enclosure movement. The difficulties of addressing questions central to the matter of environmental justice are enormous, especially today, when structural adjustment programs (SAPS), economic liberalization, and the World Trade Organization (WTO) regulations promise almost unlimited global "plunder" by transnational corporations.[18] Under IMF and World Bank structural adjustment programs, countries would seem to have little choice. Environmental issues are difficult to address in the face of the prevailing strength of the global ideological hegemony of "markets" and so-called market rationality. "In a sense, the market is the institutionalization of individualism and non-responsibility."[19]

Even so, in the 1990s grassroots movements emerged across many countries where people reasserted traditional collective and communitarian values over those of individualism.[20] This overall movement was, in part, a reaction to the cultural imperialism of capitalist development. Stripping people of cultural traditions and use value is also linked to stripping them of land. People were swept off the land in the first place, in the enclosure movement, and driven into the city where their labor would be available for the production of surplus value. Their land and labor were both commodified. Only by controlling the land can people control use value and their resources. In the face of this new capitalist enclosure, emerging grassroots movements and community organizations seek to address the real needs of communities. They can allow people to define their basic needs and develop the resources to meet these needs with minimal outside assistance.[21] These factors are closely linked to preserving and improving the environment in communities where people live.

Claire Van Zevern points to the strong ethic to care for the land among Native Hawaiians, similar to the land ethic of Aldo Leopold.[22]

Like the Native Americans, they were subjected to cultural imperialism since "blatant force" was used to compel the king to sign the "Bayonet Constitution" and hand over power to the House of Nobles, dominated by American missionaries, in 1875. With the annexation in 1898, Hawaiians began to be denied the traditional rights to the land. As Zevern points out, "The most fundamental aspect of cultural, social, economic, political, and human survival is land, without which these rights cannot be sustained."[23] As with Native Americans, in Hawaii, the "communal land base system" was destroyed. This was also the loss of culture since "cultural survival is based upon the human right to develop self-sufficient economies which require a land base."[24] The loss of the land was the loss of social power. This, in a nutshell, is what is happening to living communities around the globe, and it is what must continue to happen when commodification and environmental devastation are on the agenda of today's global corporation.

It is significant that the insights into the nature of production from thinkers such as Mahatma Gandhi and E. F. Schumacher emerged in grassroots movements in the 1990s in resistance to industrialization and "development" that destroys traditional communities. These thinkers, along with John Ruskin in England, essentially approached the problem from an anticapitalist point of view and arrived at similar conclusions about the capacity of capitalism to destroy living communities. Put simply, "the substance of man cannot be measured by gross national product."[25] Schumacher argued that the mainstream view of economists that the "problem of production" had been solved was misguided because "the earth's capital is being used up."[26] Developmentalism sends the message to the poor and marginalized peoples of the world that in due time, only if they are patient, the good life will trickle down to them also.[27] Yet the "carrot of consumerism" seems increasingly a "cruel deception" for hundreds of millions.[28] It is little wonder that people of the Global South have seen the necessity to pursue alternative paths to the "good life" that work for them. However, the corporate drive to commodify the resources of the entire globe today militates against such efforts regardless of how fervent they might be.

Big-Projectism in the Global South

In most cases, developmentalism has come to people in the form of "big-projectism." Typically, communities and their way of life, and all they own, are treated as merely an impediment to "progress" and are bulldozed out of the way to make way for the future. In the Global South, the projects that do this tend to be big dams, steel mills, the timber industry, mining, and so on. In recent years, resistance to such projects has increased dramatically. In eastern India, for example, in 1996, 500 people demonstrated against the effort of TISCO, a large steel firm, to locate a mill that would displace some 22,000 people in 27 villages. The factory would destroy forever the traditional crops such as rice, cashews, jackfruit, and the kewda plant. In this case, the area targeted produces 90 percent of India's kewda plant, which is used as an aromatic flavor and provides a livelihood for the local inhabitants. Kewda cultivation is a significant example of a livelihood in harmony with nature and the local economy. As the plant needs little investment, peasants need not borrow money; it grows on marginal land on the edge of fields, and the bushes need no watering. In addition to the valuable aromatic flavor extracted, the fibers provide a livelihood to families who make ropes, bags, and other products from the plant. Peasants organized a local group, the *Gana Sangram Samati*, to prevent government and TISCO vehicles from entering the area. Educational seminars were conducted to inform the people about the devastating impact of the proposed project.[29]

In another case, the Indian army planned to take over tribal land in eastern India for artillery practice. In such cases, much of the land eventually ends up in the hands of non-tribal people—a result which demonstrates a form of land alienation through state-sponsored projects.[30] In fact, this type of appropriation has a long history whereby the forest and mineral wealth are taken from tribal peoples. This alienation of tribal people from their ancestral land has intensified under globalization and liberalization.[31]

Perhaps the most dramatic example is the well-known project to construct a series of huge dams in the Narmada Valley in western India.[32] The project includes two large dams, the Sardar Sarovar Project (SSP) and the Narmada Sagar Project (NSP), as well as 28 other large dams. The project has displaced at least 200,000 individuals. The SSP and the NSP will submerge over 100,000 hectares of land, including 50,000 hectares of forest. The Department of Environment and Forests of the Indian government estimated in 1987 that the environmental losses due to constructing the dams would be three times the cost of the SSP.[33] It was generally acknowledged that the state governments involved did not have the resources even to resettle the displaced population, much less to address the social costs of destroying established communities. The social costs of breaking up communities by moving families to new areas and attempting to resettle families on scattered plots, rather than in an organic community, could not be taken into account.[34] Scholars and observers pointed out that there was no comprehensive rehabilitation scheme for thousands who would lose their livelihood as artisans and fisherfolk and be forced into a life of toil as agricultural and manual laborers.[35] Critics also argued that the primary beneficiaries of the project would be in the relatively prosperous and industrialized areas of Gujarat state.[36]

A massive struggle against the dam project was spearheaded by the Narmada Bachao Andolan (NBA), under Medha Patkar, who fasted and led the struggle to stop the project.[37] In 1996, the Indian Supreme Court ordered that construction on the dam be suspended. Gujarat state officials reportedly declared that the Supreme Court order "held little relevance to them."[38] States excluded people from the decision-making process, and local chambers of commerce portrayed the people as being in favor of the project.[39] As costs mounted, the benefits of such massive developmentalist schemes, which destroy forests, ecosystems, villages, communities, and livelihoods for hundreds of thousands, were questioned.[40] Resistance groups argued that development must be "people oriented."[41]

People are questioning large projects and the resulting displacement, which, it has been argued, is an objective of such projects in terms of creating a huge pool of cheap labor needed for development activities such as industry and agriculture.[42] The record shows that in 40 years, at least 18.5 million people have been displaced, 77 percent of whom are tribal people. Only 29 percent have been rehabilitated, by government estimates.[43] An NGO, the National Alliance of People's Movements (NAPM), has brought together many grassroots movements representative of the non-party activism emerging in the Third World. These groups demand "equality, simplicity, and self-reliance for proper development."[44]

NAPM sets out a number of principles for "people's development." First, the aim of development must be permanent peace, fulfillment, and happiness, not prosperity in terms of material acquisition. Second, people must have control over the natural resources in their vicinity. Land should be owned by the tiller, the commons restored to the people, and planning must involve the people in decision making. Third, natural resources should be used for the fulfillment of the needs of all, not a portion of society. Fourth, there should be self-reliance in both urban and rural communities, with a limited dependence on expanded markets. Fifth, production should be decentralized, largely based on renewable energy. Sixth, austerity and simplicity should be organizing principles, as opposed to exploitative consumerism. Seventh, alternatives to resources like metals and petroleum must be found. Eighth, there should be sustainable use and conservation of soil, waters, and forests. Ninth, living relationships should be established between producer and consumer. Tenth, production should be organized around labor-intensive techniques with the right to work guaranteed as a fundamental right. Eleventh, justice and equity should be pursued through positive and protective discrimination toward egalitarian goals. Finally, nonviolence and dialogue should be used as the proper means of change, by individuals as well as societies. The "destructive, self-defeating nature of the high-consumption model" would become self-evident.[45]

Paul Ekins explored many grassroots projects based on decentralized people's initiatives in a number of countries.[46] Among these are Feminist Centre for Information and Action (CEFEMINA) in Costa Rica. Through this organization, women help plan their communities, build their own houses, and carry out environmental actions.[47] Survival International, an international NGO to support indigenous peoples, helped prevent NATO from building a second air base on Innu land in Canada and stopped the construction of a pulp mill on tribal land in Indonesia by Scott paper, among other successful campaigns.[48] Another example of people's initiatives is the *Sarvodaya Shramadana* Movement in Sri Lanka. The movement contributed to village reconstruction in some 8,000 villages as a way of keeping local capital at home, not in the coffers of transnational corporations. In the early 1990s, the organization was active in about one third of all villages in the country.[49]

Forests for the People

Conservation and management of the forests are also important elements of environmental struggle in the Global South. In Thailand, Buddhist monks have "sanctified" trees with saffron cloths to protect the rain forests.[50] Grassroots movements to protect forests are widespread and functioning in many villages in various parts of India and other countries. The well-known Chipko movement began in India in the 1970s, with women hugging the trees to prevent the government's cutting of forests.[51] Women in many villages took over the management of forests themselves, organizing forest protection committees and stationing individuals to keep a watch over the forests to prevent damage. They also planted trees and began to manage the harvest of products from the forest. The results of this movement have been very significant. Village women have become local leaders and organized new economic activities based on forest products. Surplus fodder can be produced and sold outside the village for extra income.[52] Through these forest protection committees, run by women, the forests are regenerated. These are natural poverty prevention programs.[53]

Another issue is the right of tribal or indigenous peoples to forests. In December 1988, Chico Mendes, the rubber tappers' leader in the Amazon, was assassinated by ranching interests while fighting for the rights of Indians to the forest.[54] The recognition of the rights of indigenous peoples to land and forests is an issue in many countries today, including the U.S. In India, forests are the basis of the tribal economy. Surveys have shown that India is rapidly losing its forests, which now cover only about 10 percent of the land area. NGOs in India, such as the *Khedut Mazdoor Chetna Sangath* (Organization for the Consciousness of Peasants and Workers) are fighting for tribal rights to forests, and there has been some movement toward recognition of the rights of tribal peoples through the granting of land titles.[55] But there are conflicts today between government forestry officials and tribals over the use of the forests, with NGOs accusing the government of following the colonial policy of short-term profits and failure to respect the right of local communities to use the forests for subsistence. Activists argue that tribal peoples should be allowed access to the forests so that they will have a stake in their care and preservation. When people earn from conservation, they will invest in it.[56]

The "tragedy of the commons" paradigm has been used to argue that common property institutions cannot work. As such, this perspective has become an "ideological construct" that supports the trend toward "states and markets reducing political space." It has been used by states to legitimize seizing control of the commons on the grounds of conservation, or the greater needs of the larger society. But the model does not, in fact, demonstrate "a real tendency toward human maximizing behavior."[57] The model is used to demonstrate that people cannot act collectively. There are, however, many examples of how common property institutions are functioning on a collective basis. There is a tradition in India, for example, of informal village-level committees that were eroded by colonial rule. The patterns of use of the commons vary widely, and restraint over the use of land is related to rituals, myths, preferences, values, cultural norms, and popular conceptions of what is moral and just. Among common property institutions were wells, bathing tanks, temples, pasture lands, and so on.[58]

Many institutions in India show that common community institutions can function well if the costs to individuals of not preserving the commons outweigh the costs of maintaining it. For example, biogas plants have been established in many villages, and such institutions may also become a source of community political cohesion.[59]

The Gene Snatchers:
Transnational Corporations and the Loss of Biodiversity

Another area of NGO activity in the Third World is directed toward the rights of farmers to genetic resources and biodiversity, which are under threat from Western multinational corporations. Under Trade Related Intellectual Property Rights (TRIPS) of the World Trade Organization, indigenous plant genetic resources (PGR), seed germ plasm, is treated as the common heritage of mankind. But once bred into new varieties of seeds, the same PGRs are owned by Western multinational corporations, patented, and become a commodity that farmers must purchase. It has been reported that germ plasm imports to the U.S. have netted the U.S. economy some $70 billion from agriculture alone. Farmers risk losing the right to produce their own seed. Farmers lose control over their seeds as virtually every useful gene is privatized, and even the characteristics of the plants are patented and owned by a dozen or so multinational biotechnology corporations. NGOs are working to preserve the rights of tribal peoples, pastoralists, herbalists, and fisherfolk to the biodiversity they have preserved over centuries.[60] Local NGOs call for the recognition of community rights over individual rights for groups such as tribals. Such rights have been recognized by the highest court in Australia.[61]

Biotechnologies sometimes encourage the increased use of toxics in food production. Some new varieties of seeds being marketed in the Global South by Western seed companies are bred to be more herbicide-tolerant. With Attrazine-resistant soybeans, three times as much of the chemical can be applied without damaging the crop.[62] This tendency toward monocultures to serve the needs of global markets is a threat to sustainable agriculture and the local ecology. Local solutions can be found that benefit the country and local communities rather than render agriculture dependent upon foreign multinational corporations.[63]

The above examples of communities seeking alternative patterns of development at a time when transnational corporations are making major inroads into penetrating markets of the Global South illustrate some of the challenges ahead for environmentally sound alternative development. The struggle for actually sustainable development with local solutions is under threat by global capitalism. The following six principles could serve as guidelines for local development: First, there should be social reforms, women's education, and people's empowerment. Second, there should be a balance between industrial and agricultural activities, between urban and rural development. Third, communal ownership should be organized for the management of natural and productive resources. Fourth, technologies should be adopted that are employment-friendly and eco-friendly. Fifth, the exploitation of natural and productive resources for rampant consumerism must be avoided. Finally, a decentralized energy strategy should be adopted, that is ecologically viable and non-exploitative.[64] In linking grassroots NGOs to international environmental organizations, states can be "educated" as to the real needs and demands of the people, helping to create real people's democracy.[65] At its root, environmental justice challenges the systemic structure of developed capitalist society and the global system of discrimination.

Endnotes

Chapter One

1. Noam Chomsky, *Deterring Democracy* (New York: Hill and Wang, 1992), pp. 331–50; Robert Brenner, "The Economics of Global Turbulence," *New Left Review*, no. 229 (May-June 1998).

2. Brenner, p. 9.

3. Harold C. Barnett, *Toxic Debts and the Superfund Dilemma* (Chapel Hill, NC: University of North Carolina Press, 1994); Robert Gottlieb, ed., *Reducing Toxics: A New Approach to Policy and Industrial Decisionmaking* (Washington, DC: Island Press, 1995); Norman J. Vig and Michael E. Kraft, eds., *Environmental Policy in the 1990s* (Washington, DC: CQ Press, 1990); James P. Lester, ed., *Environmental Politics and Policy: Theories and Evidence* , 2nd ed. (Durham, NC: Duke University Press, 1995); Lamont C. Hempel, *Environmental Governance* (Covelo, CA: Island Press, 1996); Brian Keating and Dick Russell, "Inside the EPA Yesterday and Today . . . Still Hazy after All These Years," *E Magazine* (July-August 1992), pp. 31–37; Mike Davis, "Dead West: Ecocide in Marlboro Country," *New Left Review*, no. 200 (July-August 1993), pp. 49–73; Lois Marie Gibbs, *Dying from Dioxin* (Boston: South End Press, 1995).

4. Barnett, *Toxic Debts*, p. 259. In fact all landfills leak and pollute the groundwater. For example, the testimony of Dennis E. Williams, a geologist and hydrologist, before the Board of Supervisors of San Bernardino County, California, May 9, 1995: "Landfills Leak." This document can be found on the www.stopwmx web-site. More technical information on both solid and hazardous waste landfills can be found in Harry M. Freeman, ed., *Standard Handbook of Hazardous Waste Treatment and Disposal*, 2nd ed. (New York: McGraw-Hill, 1998). The fragility and temporary nature of landfills can be understood from several articles here. Even when the best technology is used, water leaks into the landfills resulting in large accumulations of hazardous leachates. Attempts are made to remove this leachate for some thirty years after the closure of the landfill, but after this, the continued accumulation will contaminate the groundwater. In discussing aboveground storage, the point is made that "In a sense, anything we do with our solid-waste residue is only storage, since it will likely remain an environmental hazard for at

least hundreds of years." Also burrowing animals go into the tops of landfills, breaching the cover. Roots of plants and trees grow through the linings. The waste in landfills may be mined in the future, presumably when environmental considerations require it or when this becomes economically feasible. Two case studies are given here. The Marathon County Area A Landfill in Ringle, Wisconsin, produced 36.71 million gallons of leachate between 1981and 1996. In 1993 a plume of contaminated groundwater was detected, containing manganese and six organic chemicals, including benzene and 1,1- dichloroethene. Technically, several problems arose. Solids in the leachate plugged up the drainage system, contaminating groundwater. Gas pressure inside the landfill resulted in numerous seeps and blowouts. Sealings around leachate removal lines could not be maintained, resulting in more contamination. Leachate depths in the cells built up to 5 to 20 feet. There were similar problems with the other facility surveyed. The engineering literature documents such problems during the active life of these landfill facilities. Once these sites are abandoned to the forces of nature, the private sector that reaped the profits has no ownership or liability for whatever happens. The public will pay the future unknown toxic debt. See K.W. Brown, David C. Anderson, and James C. Thomas, "Aboveground Disposal," in Freeman, ed., *Standard Handbook of Hazardous Waste Treatment and Disposal,* pp. 10.66–10.75.

5. Barnett, Toxic Debts, p. 170.

6. Ibid., p. 244.

7. Ibid., p. 170. Today, engineers stress that some technologies are not proven and there are many problems in the operation of landfills. See, for example, Robert Hauser, "Ash Leachate Problem Demands Innovative Solution," *Waste Age,* February 1997, wasteage.com; Cheryl L. Dunson, "Controlling Leachate When Your Juice Gets Loose," *Waste Age,* March 1997, wasteage.com; Charles W. Leung and David E. Ross, "Hazardous Waste Landfill Construction," in Freeman, ed., *Standard Handbook of Hazardous Waste Treatment and Disposal,* pp. 10.3–10.30. Many technical problems that arise are discussed here and in other articles in the same volume.

8. Barnett, *Toxic Debts,* p. 49.

9. Tom Athanasiou, "The Age of Greenwashing," *Capitalism, Nature, Socialism* 7, no. 1 (March 1996), pp. 1–36; Gregg Easterbrook, *A Moment on the Earth: The Coming of Environmental Optimism* (New York: Viking, 1995); Michael Fumento, *Science Under Siege: Balancing Technology and the Environment* (New York: William Morrow and Company, Inc., 1993).

10. John Bellamy Foster, *The Vulnerable Planet* (New York: Monthly Review Press, 1994), chap. 7.

11. Walter H. Corson, ed., *Global Ecology Handbook* (Boston: Beacon Press, 1990), p. 246.

12. Barnett, *Toxic Debts,* p. 14.

13. Maureen Smith and Robert Gottlieb, "The Chemical Industry: Structure and Function," in Gottlieb, ed., *Reducing Toxics,* pp. 210–11; U.S. Environmental Protection Agency, Office of Pollution Prevention and Toxics, *1996 Toxics Release Inventory* (Washington, DC: U.S. Government Printing Office, 1998), p. 207.

14. Smith and Gottlieb, "The Chemical Industry: Structure and Function," pp. 209–32; U.S. EPA, *1996 TRI,* pp. 207–11; Jan Naar, *Design for a Livable Planet* (New York: Harper and Row, 1990), p. 39.

15. *Statistical Abstract of the United States,* 1998 (Washington, DC: U.S. Government Printing Office, 1998), pp. 741, 874.

16. U.S. Environmental Protection Agency, Office of Pollution Prevention and Toxics, *1996 TRI*, p. 207.

17. *Statistical Abstract*, p. 751.

18. Ibid., p. 563.

19. U.S. Environmental Protection Agency, Office of Pollution Prevention and Toxics, *1996 TRI*, p. 208.

20. Ibid., pp. 207, 210.

21. Smith and Gottlieb, "The Chemical Industry: Structure and Function," p. 209.

22. U.S. Environmental Protection Agency, Office of Pollution Prevention and Toxics, *1996 TRI*, pp. 212–13.

23. Ibid., pp. 216–17.

24. Ibid., p. 218.

25. Barnett, *Toxic Debts*, p. 11.

26. Ibid.

27. MacKenzie L. Davis, "Definition and Classification of Hazardous Waste," in Freeman, ed., *Standard Handbook of Hazardous Waste Treatment and Disposal*, p. 2.3.

28. Ibid., pp. 2.3–2.5.

29. Ibid., pp. 2.5–2.7

30. Ibid., p. 2.10.

31. Ibid.

32. Ibid., p. 2.11.

33. Ibid., p. 2.12.

34. John F. Chadbourne, "Cement Kilns," in Freeman, ed., *Standard Handbook of Hazardous Waste Treatment and Disposal*, pp. 8.58, 8.76.

35. U.S. Environmental Protection Agency, Office of Pollution Prevention and Toxics, *1996 TRI*, pp. 5–6.

36. Ibid., p. 21.

37. Kristen A. Crockford, "Quantities of Hazardous Waste," in Freeman, ed., *Standard Handbook of Hazardous Waste Treatment and Disposal*, p. 2.35.

38. U.S. Environmental Protection Agency, Office of Pollution Prevention and Toxics, *1996 TRI*, pp. 11–12.

39. Ibid., pp. 9–10, 55.

40. Robert Gottlieb and Maureen Smith, "The Pollution Control System: Themes and Frameworks," in Gottlieb, ed., *Reducing Toxics*, p. 11.

41. Ann Misch, "Assessing Environmental Health Risks," in Lester, et al., *State of the World 1994* (New York: Norton, 1994), pp. 117–36; Janice Mazurek, Robert Gottlieb, and Julie Roque, "Shifting to Prevention: The Limits of Current Policy," in Gottlieb, ed., *Reducing Toxics*, p. 62. This source claims the EPA had complete data on only six chemicals by 1990.

42. Gary Cohen and John O'Connor, eds., *Fighting Toxics* (Washington, DC: Island Press, 1990), pp. 12–15, 171–73.

43. Gibbs, *Dying from Dioxin*. This is the preeminent book on dioxin. It documents important studies and information on dioxin and is also a manual on how to research and organize against dioxin producers in local communities.

44. Corson, ed., *Global Ecology Handbook*, pp. 252-53.

45. Foster, *The Vulnerable Planet*.

46. Walter Rosenbaum, "The Bureaucracy and Environmental Policy," in Lester, ed., *Environmental Politics and Policy*, p. 206. Vig and Craft, eds., *Environmental Policy in the 1990s*, Appendix I.

47. Rosenbaum, "The Bureaucracy and Environmental Policy," pp. 207–08.

48. Robert Gottlieb, Maureen Smith, and Julie Roque, "By Air, Water, and Land: The Media-Specific Approach to Toxics Policies," in Gottlieb, ed., *Reducing Toxics*, p. 27.

49. Ibid.

50. Ibid., pp. 27–33.

51. Ibid., pp. 36–37.

52. Ibid.

53. Vig and Craft, eds., *Environmental Policy in the 1990s*, Appendix I.

54. Robert Gottlieb and Janice Mazurek, "Introduction," in Gottlieb, ed., *Reducing Toxics*, p. 4.

55. Keating and Russell, "Inside the EPA," pp. 30–37.

56. Ibid., p. 37.

57. Evan J. Ringquist, "Political Control and Policy Impact in EPA's Office of Water Quality," *American Journal of Political Science* 39, no. 2 (May 1995), pp. 336–63; Barnett, *Toxic Debts*, pp. 42–49 and chap. 4.

58. Barnet, *Toxic Debts*, pp. 71–75.

59. Ringquist, "Political Control and Policy Impact in EPA's Office of Water Quality," pp. 359–60; Barnett, *Toxic Debts*, pp. 79–81.

60. Gottlieb and Smith, "The Pollution Control System," p. 17.

61. Robert Gottlieb, "The Difficulty of Getting There: The Evolution of Policy," in Gottlieb, ed., *Reducing Toxics*, p. 8.

62. Gottlieb and Smith, "The Pollution Control System," p. 19.

63. Gottlieb, "The Difficulty of Getting There," p. 9.

64. Gottlieb and Smith, "The Pollution Control System," p. 15.

65. Gottlieb, Smith and Roque, "By Air, Water, and Land," pp. 30–34.

66. Mazurek, Gottlieb, and Roque, "Shifting to Prevention," pp. 59–60.

67. Ibid., pp. 62–65.

68. Ibid., p. 68.

69. Ibid., pp. 75–76.

70. Ibid., pp. 74, 76–77.

71. Ibid., pp. 77–84.

72. James P. Lester, "Federalism and State Environmental Policy," in Lester, ed., *Environmental Politics and Policy*, p. 40. Interestingly, but not surprisingly, the concept of federalism does not exist when it comes to the use of the Commerce Clause of the U.S. Constitution, which serves as a powerful corporate tool to allow waste corporations to move waste at will across state lines to increase profits and capitalist accumulation. For example, the Association for Waste Hazardous Materials Transporters (AWHMT) points out on its website that "[o]ver the last few years, the American Truckers Association (ATA) Litigation Center has fought and won attempts by many states to impose hazardous waste and other fees on motor carriers. Thanks to ATA's efforts, millions of dollars in refunds have been court-awarded to the trucking industry." It is clear that the Commerce Clause as interpreted by the courts is a powerful instrument for increasing industry profits and capitalist accumulation in the waste industry. www.tsdcentral.com/awhmt.

73. Ibid., pp. 44–47.
74. Barnett, *Toxic Debts,* pp. 68–69; Lester, "Federalism and State Environmental Policy," in Lester, ed., *Environmental Politics and Policy*, p. 42.
75. A. Myrick Freeman III, "Economics, Incentives, and Environmental Regulation," in Vig and Craft, eds., *Environmental Policy in the 1990s*, p. 152.
76. Freeman, "Economics, Incentives, and Environmental Regulation," pp. 145–66; Lettie McSpadden, "The Courts and Environmental Policy," in James P. Lester, ed., *Environmental Politics and Policy*, pp. 242–73; Peter Montague, "The Right to Pollute," *Rachel's Environment and Health Weekly* # 442, May 18, 1995, www.monitor.net/rachelr442.html.
77. Freeman, "Economics, Incentives, and Environmental Regulation," p. 156.
78. Ibid., p. 149.
79. Ibid., p. 150.
80. Peter Montague, "Environmental Trends," *Rachel's Environment and Health Weekly*, # 613, August 27, 1998, www.monitor.net.
81. McSpadden, "The Courts and Environmental Policy," pp. 257–58; Freeman III, "Economics, Incentives, and Environmental Regulation," pp. 156–57.
82. McSpadden, "The Courts and Environmental Policy," pp. 257–58.
83. Ibid.
84. Freeman, "Economics, Incentives, and Environmental Regulations," p. 159.
85. McSpadden, "The Courts and Environmental Policy," pp. 259–60.
86. Barnett, *Toxic Debts*, p. xiii.
87. Ibid., p. 2.
88. Ibid., pp. 23–24.
89. Ibid., pp. 24–25.
90. Ibid., p. 28.
91. Ibid., p. 64.
92. Ibid., p. 74.
93. Ibid., pp. 74–75.
94. Ibid., p. 41.
95. Ibid., pp. 212–19.
96. Ibid., p. 230.
97. Ibid., p. 257.
98. Ibid., p. 259.
99. Ibid., pp. 263–64.
100. Ibid., pp. 265–66.
101. Athanasiou, "The Age of Greenwashing"; Gene M. Grossman and Alan B. Krueger, "Economic Growth and the Environment," *Quarterly Journal of Economics* 110, 2 (May 1995), pp. 353–77; Easterbrook, *A Moment on the Earth;* Merle Jacob, "Toward a Methodological Critique of Sustainable Development," *Journal of Developing Areas* 28, 2 (January 1994), pp. 237–52.
102. Athanasiou, "The Age of Greenwashing," 32–33.
103. Merle Jacob, "Toward a Methodological Critique of Sustainable Development," pp. 241–43.
104. Ibid., p. 245.
105. Ibid., p. 246.

106. Ibid., pp. 246–52.

107. Athanasiou, "The Age of Greenwashing," pp. 30–31.

108. Ibid., pp. 31–32.

109. Ibid., pp. 1-2.

110. Keating and Russell, "Inside the EPA," p. 35.

111. Athanasiou, "The Age of Greenwashing," p. 6.

112. Keating and Russell, "Inside the EPA," p. 37.

113. Athanasiou, "The Age of Greenwashing," pp. 12–18.

114. Lamont C. Hempel, *Environmental Governance* (Covelo, CA: Island Press, 1996), p. 119.

115. Gregg Easterbrook, "The Illusions of Scientific Certainty," in Russ Hoyle. ed., *Gale Environmental Almanac* (Detroit: Gale Research, 1993), pp. 307–360.

116. For example, see the discussion in Lois Gibbs, *Dying from Dioxin* (Boston: South End Press, 1995), Introduction, pp. 1–20.

117. Easterbrook, "The Illusions of Scientific Certainty," p. 308.

118. Michael Fumento, *Science Under Siege: Balancing Technology and the Environment* (New York: William Morrow and Company, 1993), pp. 367–69.

119. Ibid., pp. 371–72.

120. Keating and Russell, "Inside the EPA," p. 34.

Chapter Two

1. "We must note once again that the fundamental problem is the unfettered power of the modern corporation," states Peter Montague, in "The Right to Know Nothing," *Rachel's Environment and Health Weekly, # 552*, 26 June 1997, www.monitor.net.

2. Karl Marx, *Capital*, vol. I, in Eugene Kamenka, ed., *The Portable Karl Marx* (New York: Penguin, 1983), pp. 465–78.

3. Stephen Horton, "Value, Waste and the Built Environment: A Marxist Analysis," *Capitalism, Nature, Socialism* 8, no. 1 (March 1997), pp. 127–39.

4. From Vance Oakley Packard, *The Waste Makers* (New York: D. McKay Co., 1960), quoted in Peter Cant, "Environmental Governance," unpublished paper (1996).

5. Marx, *Capital*, vol. I, in Kamenka, ed., *The Portable Karl Marx*, p. 495.

6. Robert Brenner, "The Economics of Global Turbulence," *New Left Review*, no. 229 (May-June 1998). Entire issue.

7. Ralph Miliband, *Divided Societies: Class Struggles in Contemporary Capitalism* (Oxford: Clarendon Press, 1989).

8. John T. Aquino, "The New Republic: One Billion Strong and Growing," *Waste Age*, October 1998, www.wasteage.com.

9. "Environmental Technologies," *U.S. Industry and Trade Outlook '98* (New York: DRI/McGraw Hill, 1998), pp. 20–2. *Statistical Abstract of the United States*, 1998 (Washington, DC: U.S. Government Printing Office, 1998), pp. 246–47.

10. Bethany Barber and John Aquino, "Fourth Annual Waste Age 100," *Waste Age*, September 1997, www.wasteage.com.

11. Paul M. Sweezy, *The Theory of Capitalist Development* (London: Dennis Dobson, 1942), p. 254.

12. Ibid., p. 257.

13. Ibid., p. 262.

14. Ibid., p. 263.

15. Ibid., p. 264.

16. *"Cumberland Farms et. al. v. Browning Ferris Industries, Inc. and Subsidiaries*: Plaintiff's Memorandum in Opposition to Defendant's Motion for Summary Judgement, 87-3717," www.enviroweb.org.

17. Sweezy, *The Theory of Capitalist Development*, p. 265.

18. Ibid., p. 272.

19. Ibid., pp. 274–75.

20. John T. Aquino, "Casella Climbs Every Mountain," *Waste Age*, April 1998, www.wasteage.com.

21. Sarah Halsted, "A Shrinking Middle Class?" *Waste Age*, May 1998, www.wasteage.com.

22. Lynn Merrill, "The Garbage STOPS Here," *Waste Age*, January 1988, www.wasteage.com.

23. Bethany Barber, "Apples vs. Oranges: Managing Public/Private Competition," *Waste Age*, October 1998, www.wasteage.com; Laith B. Ezzet, "Collection: Who Handles the Trash in the 100 Largest Cities," *Waste Age*, April 1998, www.wasteage.com.

24. Barber, "Apples vs. Oranges."

25. Tom Kerr, "Developing Landfill Gas Projects without Getting Credit," *Waste Age*, May 1998, www.wasteage.com.

26. Jerry Ackerman, "Turning Brownfields into Green," *Waste Age*, May 1998, www.wasteage.com; "PA, NJ Agencies Empowered by Brownfield Laws," 19 April 1999. www.solidwaste.com.

27. "PA, NJ Agencies Empowered by Brownfield Laws." It is interesting to note that the governor who signed the Brownfield law in New Jersey in 1998 was Christie Whitman, now head of the EPA in the Bush administration. Recently, President Bush signed legislation increasing brownfields funding, with states and localities eligible to receive up to $250 million annually for five years to clean up polluted sites; potential developers have no liability for cleaning up toxic wastes existing at a site prior to the time of purchase. (See Lew Sichelman, "Feds Increase Brown-fields Funding," *Realty Times*, 14 January 2002, http://realtytimes.com/rtnews/rtcpages/20020114_brownfields.htm.) Liability is indeed a crucial issue in pro-tecting corporate assets. Under Michigan law, for example, stockholders and par-ent corporations are shielded from liability. Corporations can spin off hazardous waste and other high potential liability operations. That is, the parent company creates a hazardous waste company as a subsidiary. This provides a great deal of protection of the parent company from potential liability. The case of *Donahey v. Bogle* in 1997, decided in the U.S. Court of Appeals for the Sixth Circuit, sheds light on why brownfields legislation is needed by companies to launder toxic lia-bilities. Bogle's brother was the sole stockholder of St. Clair Rubber Company in Marysville, Michigan. Donahey later bought the property for $115,000 with St. Clair agreeing to pay for the cost of cleanup. Instead, St. Clair simply ceased to exist as a corporation after the sale. St. Clair had polluted the property so badly that it would cost $1 million to clean it up. The court ruled that Bogle was stuck with the toxic liability, with the former sole shareholder having no liability under the Superfund law. The court relied primarily on the case *of U.S. v. Cordova Chem*

Co., 113F.3d 572. See Barry Shanoff, "Legislation Clears Stockholder in Cleanup Suit," *Waste Age,* March 1998, www.wasteage.com.

28. Jerry Ackerman, "Turning Brownfields into Green."

29. Peter Montague, "Incineration News," *Rachel's Environment and Health Weekly,* # 592, 2 April 1998, www.monitor.net.

30. Charles E. Faupel, Conner Bailey, and Gary Griffin, "Local Media Roles in Defining Hazardous Waste as a Social Problem: The Case of Sumter County, Alabama," *Sociological Spectrum* 11 (1991), pp. 293–319; Matthew C. Urie, "Share and Share Alike? Natural Resources and Hazardous Waste under the Commerce Clause," *Natural Resources Journal* 35, no. 2 (Spring 1995), pp. 309–80. The practice of dumping toxics in black neighborhoods is widely documented in the environmental justice literature. See Robert D. Bullard, *Dumping in Dixie: Race, Class and Environmental Quality* (Boulder: Westview Press, 1990); Robert D. Bullard, *Unequal Protection: Environmental Justice and Communities of Color* (San Francisco: Sierra Club Books, 1994); Robert Bullard, "Environmental Blackmail in Minority Communities," in Bunyan Bryant and Paul Mohai, eds., *Race and the Incidence of Environmental Hazards: A Time for Discourse* (Boulder: Westview Press, 1992), pp. 82–95; Robert D. Bullard, ed., *Confronting Environmental Racism: Voices from the Grassroots* (Boston: South End Press, 1993); Francis O. Adeola, "Environmental Hazards, Health, and Racial Inequality in Hazardous Waste Distribution," *Environment and Behavior* 26, no. 1 (January 1994), pp. 99–126; Harvey L. White, "Hazardous Waste Incineration and Minority Communities," in Bryant and Mohai, eds., *Race and the Incidence of Environmental Hazards,* pp. 126–39.

31. Urie, "Share and Share Alike?" pp. 361–62.

32. Bullard, *Unequal Protection.*

33. Faupel, Bailey, and Griffin, "Local Media Roles," p. 305.

34. Conner Bailey and Charles E. Faupel, "Environmentalism and Civil Rights in Sumter County, Alabama," in Bryant and Mohai, eds., *Race and the Incidence of Environmental Hazards,* p. 147.

35. Ibid., pp. 148–49.

36. Faupel, Bailey, and Griffin, "Local Media Roles," pp. 293–319.

37. Ibid., p. 313.

38. Urie, "Share and Share Alike?" pp. 361–63.

39. Ibid. See cases discussed here.

40. Ibid., p. 363.

41. Ibid., pp. 363–67.

42. Ibid., p. 365.

43. Ibid.

44. Ibid., p. 366.

45. Ibid., p. 367.

46. Ibid., p. 371.

47. John F. Chadbourne, "Cement Kilns," in Harry M. Freeman, ed., *Standard Handbook of Hazardous Waste Treatment and Disposal* (New York: McGraw-Hill, 1998), pp. 8.56–8.58, 8.76. Many de facto hazardous wastes are being used to produce new products. Fly ash, which collects in the flues of incinerators, is used to produce a lightweight aggregate (LWA) for the construction industry in

Wisconsin. The lightweight, ceramic-like aggregate in pellets about one-half inch in diameter is used in multi-story structures such as hospitals, libraries, and office buildings. The product is made from a mixture of fly ash from an electric plant, wastewater sludge from the Milwaukee Metropolitan Sewerage District, and paper mill sludge from southern Wisconsin paper mills. After these materials are mixed and pelletized, they are burned in a rotary kiln at 2,000 degrees F. The organic and volatile materials are to be burned out. In other words, they are put through a hazardous waste incinerator, a process that produces the ceramic-like aggregate. This is not a small operation. Some 90,000 tons per year of the material is being produced. The materials used in the production would normally go to landfills. In a way, it seems, hazardous waste incinerators, due to public opposition, have been redefined as manufacturing plants. Such operations are really a form of sham recycling, just using toxic wastes for fuel and building materials. It is not really certain if this process is better than landfilling these toxic materials. See Mary Carpenter, "Recycling Fly Ash and Sludge for Profit," *World Wastes Online*, March 1997, www.wasteage.com.

48. Chadbourne, "Cement Kilns," p. 8.56. Liquid hazardous waste used as fuel is exempted from the handling and disposal requirements of the substance as a hazardous waste.

49. Ibid.

50. Ibid., pp. 8.69–8.70.

51. Ibid., p. 8.59.

52. Robert Nozick, *Ideology, State, and Utopia* (New York: Basic Books, 1974).

53. Chaundra Frierson, "Californians Recycle Record Amount of Waste: Recyclers Keep 100 Million Tons from the Landfill," *Waste Age*, www.wasteage.com

54. Sanford Lewis, "The Corporate Right to Cover Up," *Multinational Monitor* 19, no. 5 (May 1998), EBSCOhost MasterFILE Elite, Item Number 890762. See also "Audit Privilege and Immunity Scorecard," http:gnp.enviroweb.org/score.html.

55. Peter Montague, "Right to Know Nothing," *Rachel's Environment and Health Weekly*, # 552, 26 June 1997, www.monitor.net.

56. Michael Tanzer, "Globalizing the Economy," *Monthly Review* 47, no. 4 (September 1995), pp. 1–15; Dave Broad, "Globalization Versus Labor," *Monthly Review* 47, no. 7 (December 1995), pp. 20–31.

57. Kristin Schafer, *What Works, Report No. 2: Local Solutions to Toxic Pollution* (Washington, DC: The Environmental Exchange, 1993). The struggle in Mercer County is reviewed here, pp. 97–99.

Chapter Three

1. Penny Newman, a special education teacher in California, worked as an organizer for the Citizens' Clearinghouse for Hazardous Waste (CCHW). See "Killing Legally with Toxic Waste: Women and the Environment in the United States," *Development Dialogue*, nos. 1 & 2 (1992), pp. 50–70.

2. Quoted in Robert Bullard, ed., *Confronting Environmental Racism: Voices from the Grassroots* (Boston: South End Press, 1993), p. 18; Newman, "Killing Legally," pp. 59–60.

3. Bullard, ed., *Confronting Environmental Racism*, pp. 9–13.

4. Ibid., p. 9.

5. Richard Hofrichter, ed., *Toxic Struggles: The Theory and Practice of Environmental Justice* (Philadelphia: New Society Publishers, 1993), p. 2. In the Third World, efforts often revolve around struggles to save and replenish the forests, to save one's own livelihood against encroachments of new industries such as steel plants, and to keep out megaprojects such as dams that destroy hundreds of villages. While marginalized, the poor often enjoy a community until uprooted. Afterwards, this is often destroyed.

6. Richard Moore and Louis Head, "Acknowledging the Past, Confronting the Present: Environmental Justice in the 1990s," in Hofrichter, ed., *Toxic Struggles*, pp. 118–27.

7. Robert D. Bullard, "Anatomy of Environmental Racism and the Environmental Justice Movement," in Bullard, ed., *Confronting Environmental Racism.*

8. Richard Moore and Louis Head, "Acknowledging the Past, Confronting the Present," pp. 119–20; Bunyan Bryant, ed., *Environmental Justice: Issues, Policies, and Solutions* (Washington, DC: Island Press, 1995), pp. 4–5.

9. Robert Gottlieb, *Forcing the Spring: The Transformation of the American Environmental Movement* (Washington, DC: Island Press, 1993), pp. 253–69.

10. Robert D. Bullard, *Dumping in Dixie: Race, Class and Environmental Quality* (Boulder, CO: Westview Press, 1990); Bullard, *Confronting Environmental Racism*; Robert D. Bullard, *Unequal Protection: Environmental Justice and Communities of Color* (San Francisco: Sierra Club, 1994); Bunyan Bryant and Paul Mohai, eds., *Race and the Incidence of Environmental Hazards: A Time for Discourse* (Boulder, CO: Westview Press, 1992); Hofrichter, ed., *Toxic Struggles*; Bryant, ed., *Environmental Justice*; Al Gedicks, *The New Resource Wars: Native and Environmental Struggles Against Multinational Corporations* (Boston: South End Press, 1993); Rick Whaley and Walter Bresette, *Walleye Warriors: An Effective Alliance Against Racism and for the Earth* (Philadelphia: New Society Publishers, 1994); Gottlieb, *Forcing the Spring*; Ward Churchill and Winona LaDuke, "Native North America: The Political Economy of Radioactive Colonialism," in M. Annette James, ed., *The State of Native America: Genocide, Colonization, and Resistance* (Boston: South End Press, 1992).

11. Parallel movements are emerging in many Third World countries based on a proliferation of new, usually locally based, nongovernmental organizations (NGOs). There is a burgeoning literature on these movements. See Thomas Princen and Matthias Finger, *Environmental NGOs in World Politics: Linking the Local and the Global* (New York: Routledge, 1994); Paul Ekins, *A New World Order: Grassroots Movements for Global Change* (New York: Routledge, 1992); Winin Pereira and Jeremy Seabrook, *Asking the Earth: Farms, Forests and Survival in India* (London: 1990); N. Patrick Peritore, "Environmental Attitudes of Indian Elites: Challenging Western Postmodernist Models," *Asian Survey* 33, no. 8 (August 1993), pp. 804–18; Subir Sinha and Ronald Herring, "Common Property, Collective Action and Ecology," *Economic and Political Weekly*, 28, nos. 27 & 28 (July 3-10, 1993) pp. 1425–32.

12. Lynn Merrill, "The Garbage STOPS Here," *Waste Age*, January 1998, www.wasteage.com; Kim A. O'Connell, "The Closure of Fresh Kills: What's the Forecast?" *Waste Age*, October 1998, www.wasteage.com; Andrea Bernstein, "D'Amato's Environmental Makeover," *The Nation*, February 24, 1997, pp. 11–16; Peter Montague, "Incineration News," *Rachel's Environment & Health Weekly*, # 592, April 2, 1998, www.monitor.net; Peter Montague, "Philadelphia Dumps on

the Poor," *Rachel's Environment & Health Weekly*, # 595, April 23, 1998, www.monitor.net.

13. Charlie Cray, *Waste Management, Inc.: An Encyclopedia of Environmental Crimes and Other Misdeeds, 1991*, www.enviroweb.org/stopwmx.

14. Peter Montague, "Right to Know Nothing," *Rachel's Environment and Health Weekly* # 552, June 26, 1997, www.monitor. net.

15. Celene Krauss, "Blue-Collar Women and Toxic-Waste Protests: The Process of Politicization," in Hofrichter, ed., *Toxic Struggles*, pp. 107–17. See also Claire McAdams, "Gender, Class and Race in Environmental Activism: Local Response to a Multinational Corporation's Land Development Plans," in Jennifer Turpin and Lois Ann Lorentzen, eds., *The Gendered New World Order: Militarism, Development and the Environment* (London: Routledge, 1996), pp. 51–69.

16. Sinha and Herring, "Common Property, Collective Action and Ecology," 1431.

17. Kristin Schafer, et. al., *What Works: Local Solutions to Toxic Pollution* (Washington, DC: The Environmental Exchange, 1993).

18. Krauss, "Blue-Collar Women," p. 109.

19. Lois Gibbs, "Forward," in Hofrichter, ed., *Toxic Struggles*, p. ix.

20. Krauss, "Blue-Collar Women," p. 109.

21. The enormous Three Gorges dam project on the Yangtze River in China was expected to uproot more than one million people. See Dunu Roy, "Large Projects: For Whose Benefit," *Economic and Political Weekly* 29, no. 50 (Dec. 10, 1994), p. 3129.

22. Noam Chomsky, *Deterring Democracy* (New York: Hill and Wang, 1992), p. 253.

23. Gedicks, *The New Resource Wars*, pp. 101–02.

24. Richard Hofrichter, "Cultural Activism and Environmental Justice," in Hofrichter, ed., *Toxic Struggles*, pp. 85–96; Sinha and Herring, "Common Property, Collective Action and Ecology," pp. 1425–32. Much marginal land in Cyprus still serves as a commons for the grazing of sheep. This is true in many countries.

25. See, for example, Robert Wade, *Village Republics* (Cambridge: Cambridge University Press, 1988) and Vandana Shiva, *Staying Alive: Women, Ecology and Survival in India* (New York: St. Martin's Press, 1989).

26. Hofrichter, "Cultural Activism and Environmental Justice," pp. 85–96.

27. Ibid.; E. F. Schumacher, *Small Is Beautiful: Economics as if People Mattered* (New York: Perennial Library, 1975), pp. 115–16.

28. Krauss, "Blue-Collar Women," p. 111.

29. Ibid., p. 113. See also Lorraine Elliot, "Women, Gender, Feminism, and the Environment," in Turpin and Lorentzen, eds., *The Gendered New World Order*, pp. 13–34. Elliot, along with Penny Newman in her article "Killing Legally," clearly stresses that the issue of environmental degradation and toxics is "gender specific" at both the global and local level. Women and children are "differentially situated" in terms of environmental degradation.

30. Activists have debated the question as to why women are more active—that is, whether it is because they are women or because they have more to lose when the village forests or commons are destroyed. See Sinha and Herring, "Common Property." See also Lorraine Elliot, "Women, Gender, Feminism and the Environment," and Claire McAdams, "Gender, Class and Race in Environmental Activism," pp. 13–34, 51–69.

31. Ynestra King, "Feminism and Ecology," in Hofrichter, ed., *Toxic Struggles*, p. 81.

32. Barbara Epstein, "Ecofeminism and Grass-roots Environmentalism in the United States"; Celene Krauss, "Blue Collar Women," pp. 148–49, 107.

33. Gottlieb, *Forcing the Spring*, pp. 274–76.

34. See Peter Montague, "Activist Mom Wins Goldman Prize," *Rachel's Environment and Health Weekly # 542*, April 17, 1997. The East Liverpool, Ohio, incinerator was initiated in 1982 by Jackson Stephens of Stephens, Inc., in Little Rock, Arkansas. Stephens was a wealthy political contributor to President Clinton. Both Clinton and Al Gore visited East Liverpool during the presidential campaign in 1992, and Clinton promised that if he was elected, the incinerator would never operate. But it was soon to operate under the Clinton-Gore "environmentalist" administration; Jan Naar, *Design for a Livable Planet* (New York: Harper and Row, 1990), pp. 40–41.

35. Janet Reitman, "Ms. West Goes to Tokyo," *Life*, October 1998, pp. 98–106.

36. Gottlieb, *Forcing the Spring*, p. 211.

37. Ibid., pp. 211–12.

38. Robert D. Bullard, "Anatomy of Environmental Racism," in Hofrichter, ed., *Toxic Struggles*, pp. 25–35; Paul Mohai and Bunyan Bryant, "Environmental Racism: Reviewing the Evidence," in Bryant and Mohai, eds., *Race and the Incidence of Environmental Hazards*, pp. 163–76; Bullard, ed., *Confronting Environmental Racism*; Harvey L. White, "Hazardous Waste Incineration and Minority Communities," in Bryant and Mohai, eds., *Race and the Incidence of Environmental Hazards*, pp. 126–39.

39. Bryant, ed., *Environmental Justice*, p. 5.

40. Rebecca A. Head, "Health-Based Standards: What Role in Environmental Justice?" in Bryant, ed., *Environmental Justice*, p. 52; Marianne Lavelle and Marcia A. Coyle, "Unequal Protection: The Racial Divide in Environmental Law," in Hofrichter, ed., *Toxic Struggles*, pp. 136–43.

41. Bullard, "Anatomy of Environmental Racism," in Hofrichter, ed., *Toxic Struggles*, pp. 26–27. See also Commission for Racial Justice, United Church of Christ, *Toxic Wastes and Race in the United States*; Charles Lee, "Toxic Waste and Race in the United States," in Bryant and Mohai, eds., *Race and the Incidence of Environmental Hazards*, pp.10–27.

42. Commission for Racial Justice, United Church of Christ, *Toxic Wastes and Race In The United States*. See also Beverly Hendrix Wright and Robert D. Bullard, "The Effects of Occupational Injury, Illness and Disease on the Health Status of Black Americans: A Review," in Hofrichter, ed., *Toxic Struggles*, pp. 153–54; Robert Gottlieb, *Forcing the Spring*, chap. 8.

43. Wright and Bullard, "The Effects of Occupational Injury," pp. 153–62; Gottlieb, *Forcing the Spring*, chap. 8.

44. Wright and Bullard, "The Effects of Occupational Injury," p. 154.

45. Beverly H. Wright, Pat Bryant, and Robert D. Bullard, "Coping with Poisons in Cancer Alley," in Bullard, ed., *Unequal Protection*, pp. 110-29.

46. Ibid., p. 111.

47. Whaley and Bresette, *Walleye Warriors*, p. xii. This can be compared to the way land has been taken from indigenous people in India and elsewhere, sometimes under colonialism, but often by the state for megaprojects. Much property taken from local people then ends up in private hands.

48. Ibid., p. 13.

49. Ibid., p. 18.

50. Ibid., p. 20.
51. Gedicks, *The New Resource Wars*, p. 190.
52. Whaley and Bresette, *Walleye Warriors*, p. 20.
53. Gedicks, *The New Resource Wars*, p. 192.
54. Whaley and Bresette, *Walleye Warriors*, p. 75.
55. Ward Churchill and Winona LaDuke, "Native North America: The Political Economy of Radioactive Colonialism," in James, ed., *The State of Native America*, p. 241; Winona LaDuke, "A Society Based on Conquest Cannot Be Sustained: Native Peoples and the Environmental Crisis," in Hofrichter, ed., *Toxic Struggles*, pp. 98–106.
56. Whaley and Bresette, *Walleye Warriors*, p. 181.
57. Ibid., p. 189.
58. Ibid., p. 193.
59. Gedicks, *The New Resource Wars*, p. 64.
60. Whaley and Bresette, *Walleye Warriors*, p. 235.
61. Joni Seager, "Creating a Culture of Destruction: Gender, Militarism, and the Environment," in Hofrichter, ed., *Toxic Struggles*, p. 59.
62. Gedicks, *The New Resource Wars*, pp. 39–40.
63. Churchill and LaDuke, "Native North America," p. 244.
64. Churchill and LaDuke, "Native North America," p. 247; Wm. Paul Robinson, "Uranium Production and Its Effects on Navajo Communities Along the Rio Puerco in Western New Mexico," in Bryant and Mohai, eds., *Race and the Incidence of Environmental Hazards*, pp. 153–62.
65. Churchill and LaDuke, "Native North America," pp. 248–49.
66. LaDuke, "A Society Based on Conquest Cannot Be Sustained," pp. 98–106.
67. John Bellamy Foster, "Ecology and Human Freedom," *Monthly Review* 47, no. 6 (November 1995), p. 22. See also, John Bellamy Foster, "Marx's Theory of Ecological Sustainability as a Nature-Imposed Necessity for Human Production," *Organization and Environment* 10, no. 3 (1997), pp. 30–31.
68. Hofrichter, *Toxic Struggles*, p. 5.
69. F. A. Hayek, *Road to Serfdom* (London: Routledge, 1986); Francis Fukuyama, *The End of History and the Last Man* (New York: Free Press, 1992).
70. Daniel Faber and James O'Conner, "Capitalism and the Crisis of Environmentalism," in Hofrichter, ed., *Toxic Struggles*, p. 16.
71. Vernice D. Miller, "Building on Our Past, Planning Our Future," in Hofrichter, ed., *Toxic Struggles*, pp. 128–35.
72. Bullard, ed., *Confronting Environmental Racism*, p. 13.

Chapter Four

1. Waste-Tech began as Chemical Treatment Services, Inc., in 1984, quickly changed its name to Waste-Tech Services, Inc., was purchased by Bechtel National, Inc., in 1985, and finally by Amoco in 1988. During this time the company successfully sited an incinerator in Lake Charles, Louisiana, in 1988. A second permit was issued for an incinerator in Kimball, Nebraska. The company made test burns in Lima, Ohio, in 1988, but this project was canceled. In the same year Amoco Waste-Tech was cited for a number of compliance problems and "lost" inspection records

at its Lake Charles location. Ecova Corporation was later set up for the sole purpose of building the Kimball facility, which later came on line. In 1999, the facility was owned and operated by Clean Harbors of Braintree, Inc. For attempts to site on Indian lands, see Rachel's, *Hazardous Waste News #239*, 26 June 1991.

2. Troy, in Lincoln County; Rockport, in Atchison County; Trenton, in Grundy County, 20 miles south of the Waste-Tech site in Mercer County.

3. Two other Waste-Tech sites were border counties, Atchison and Kimball, Nebraska.

4. *Commercial Atlas*, 123rd ed. (Rand McNally, 1992), pp. 395–96.

5. In 1990, Premium Standard Farms was just gaining a foothold in the area. This company was soon to become a huge intensive hog farming operation spreading over several counties. Today, under ContiGroup Companies, Inc., it is the major employer in the area.

6. "Taylor Says Development Board Talking to Number of Industrial Prospects," *Princeton Post-Telegraph* (hereinafter *PT*), 15 March 1990. One item of dispute during the whole Waste-Tech struggle was whether or not Waste-Tech intended to site an incinerator in Mercer County. The *St. Joseph (Mo.) News-Press Gazette* (hereinafter *NPG*) reported on 12 March 1990 that a spokesperson for the Department of Natural Resources stated that "preliminary meetings with Waste-Tech and members of the local board focused on a waste incinerator and landfill." Mercer County citizens believed that Waste-Tech did intend to build an incinerator as part of their project since they had plans to build them countrywide, and they had obtained options on 900 acres of land when their landfill project called for only 175. See "Waste Tech Makes It Categorical: No Incinerator in Mercer County," *PT*, 8 November 1990. See also "Waste-Tech Has Plans to Build Incinerators Across the Country but Not Mercer County," *NPG*, 22 November 1990. Waste-Tech had attempted to site an incinerator in Trenton, Missouri, just prior to moving into Mercer County. Two other incinerator sites were planned for Kimball, Nebraska, and Madison County, Florida. (Due to plenty of opposition in Madison County, Waste-Tech pulled its application for an incinerator and landfill there on October 12, 1993.) Waste-Tech, nonetheless, continued to deny planning an incinerator for Mercer County, close to the time the company pulled its application. See, for instance, "Tate Calls for DNR to Consider Nestlé Report," *PT*, 30 July 1992.

7. "Taylor Says Development Board Talking to Number of Industrial Prospects," *PT*, 15 March 1990.

8. Ibid.

9. *PT*, 5 July 1990.

10. Quoted in "Commissioners Sign Resolution Against Hazardous Waste Park," *PT*, 28 June 1990.

11. "Commissioners Sign Resolution."

12. Personal interview with Ed King.

13. "Peg's Ponderings," *PT*, 28 June 1990; "Letters Represent Comments Received," PT, 9 August 1990.

14. Sometimes regional media coverage was counterproductive. For instance, Channel 3, Kirksville, reported the Washington Township Planning and Zoning Commission in favor of the Waste-Tech project, outraging several local citizens.

15. The CCE committees included: Agriculture; Newsletter; Education and Communication; and Political, Legal, and Zoning.

16. "CCE, Mercer County PAC Appears Non-compliant with Amended Campaign Finance Disclosure Law," *PT*, 9 August 1990.
17. In Mercer County, Taylor received 471 votes, with the incumbent receiving 1,247. See "Mercer County Election Results," PT, 8 November 1990.
18. Evidence of the political heat caused by the Waste-Tech issue is clear in the results of the November 1990 election results in Mercer County, with Coleman receiving six votes less than a local opponent of Waste-Tech, a political "unknown." See "Mercer County Election Results," *PT*, 8 November 1990.
19. Rabe, *Beyond Nimby*, pp. 3, 28–33. A third, but much less frequent approach, is the voluntary approach., which Rabe argues for, pp. 4–6. For an extensive discussion of preemption and compensation, see Duffy, "State Hazardous Waste Facility Siting," 755–804. The Massachusetts "compensation" model has been a subject of much discussion in the literature. Besides Rabe's own discussion of its failure, pp. 34–38, see Michael O'Hare and Debra Sanderson, "Facility Siting and Compensation," pp. 364–76. For an apology for a more inclusive concept of compensation, an "integrated development plan" that will leave "all residents better off," see Patrick Field, Howard Raiffa, and Lawrence Susskind, "Risk and Justice: Rethinking the Concept of Compensation," *The Annals of the American Academy*, 545 (May 1996), pp. 156-64.
20. See Barry George Rabe, *Beyond Nimby*, pp. 12–13; Dennis M. Toft, "Site Selection for Hazardous Waste Facilities," *NR&E* 7, no. 3 (Winter 1993), p. 6.
21. Toft, "Site Selection," 7, mentions Massachusetts, which is a famous example of this method. For a more comprehensive discussion of the Massachusetts model, see Barry G. Rabe, *Beyond Nimby*, pp. 34–38. For state laws denying local governments "the right to prohibit or unduly restrict" the construction or operation of hazardous waste facilities, see Duffy, "State Hazardous Waste Facility Siting," pp. 792–93.
22. Neil R. Shortlidge and S. Mark White, "Use of Zoning and Other Local Controls for Siting Solid and Hazardous Waste Facilities," *NR&E*, p. 4.
23. Ibid., p. 5.
24. Ibid., p. 3.
25. A piece in *PT* 12 July 1990, stated: "The law states that the commissioners cannot pass an ordinance or law concerning the location of a hazardous waste facility in the county." The August 2, 1990, editorial argued that "there is a legal question as to whether or not they can even ATTEMPT to derail [the project]." The writer proceeded to quote the law referencing the "exception of local option on location." One wonders whether these articles were simple misunderstanding or outright misinformation.
26. This is not true for solid waste facilities. See Section 260.215.2 RSMo., which states: "Nothing in sections 260.200 to 260.245 shall usurp the right of a city or county from adopting and enforcing local ordinances, rules, regulations, or standards... equal to or more stringent than the rules or regulations adopted by the department pursuant to sections 260.200 to 260.245." Madison Township, with stringent solid waste regulations, recently stopped an 11-county regional dump being sited in their township.
27. CCE newsletters (untitled) of October 1990, July 1991, July 1992.
28. This was one body responsible for both planning and zoning.
29. The commission decided on a minimum surety bond of $35,000 to gain public support on this divisive issue. At the public hearing on June 18, 1991, the com-

mission took a vote on the surety bond issue. The bond received a tie vote (13 in favor, 13 opposed). A vote was also taken on setback distances. A one-half mile setback distance received the most votes. In 1997, the Missouri Supreme Court ruled for Premium Standard Farms against Lincoln Township, Putnam County, finding that the township's zoning regulations "exceed[ed] the township's statutorily granted zoning powers." This ruling gutted zoning restrictions against corporate hog farms—treating huge industrial operations as regular "farming," which cannot be restricted by zoning in Missouri.

30. Letter to G. Tracy Mehan III, Director, Missouri Department of Natural Resources, November 1, 1991.

31. A "residential" and "other" land use districts were created as "floating zones," or potential land use districts.

32. Washington Township Zoning Regulations, Section 501.1

33. According to James G. Trimble, Kansas City environmental lawyer, one problem with the Schuyler County ordinance—and thus with the ordinances modeled after this one, including the Washington Township one—was the lack of criteria for conditional use permits, which pertain to hazardous and solid waste facilities.

34. Washington Township Zoning Regulations, Section 905.0. One provision of the Washington Township ordinance includes the position of director, to be funded by a yearly filing fee of $100,000.

35. Supra, note 6.

36. Suzanne Jones, *NPG*, 20 June 1991.

37. A case handled by James G. Trimble, who represented both Washington and Madison townships.

38. Letter to G. Tracy Mehan III.

39. Steve Danner, "Senate Bill 420—An Attempt at a Consensus," unpublished manuscript, p. 1.

40. Suzanne Jones, "Mercer County to Vote on Ash Landfill Proposal," *NPG*, 6 February 1991.

41. "Mercer County to Express Opinion on Hazardous Waste in April Election," *PT*, 6 February 1991.

42. Opinion No. 85-91, 21 February 1991.

43. In the first case, Missouri Attorney General Opinion No. 478, Weir, 1969, it was stated that Jefferson County could not place a referendum on the ballot on the question of the termination of planning and zoning. In the second case, Missouri Attorney General Opinion Letter No. 111, Harper, 1978, the question was whether voters could give a yes or no opinion on the construction of the Meramec Dam and Lake. The third case, Missouri Attorney General Opinion No. 204-87, involved the issue of whether a referendum could be held on the sale or lease of Boone County Hospital.

44. Steve Danner, "Senate Bill 420," p. 2.

45. Suzanne Jones, "Hearing on Ballot Request Tuesday," *NPG*, 23 March 1991; Hugh F. O'Donnell, III, "The System and the Environment: Another Example of the Erosion of Democracy," unpublished manuscript, p. 3.

46. O'Donnell, "The System and the Environment," p. 4.

47. "Judge finds Mercer County Not Authorized to Conduct Referendum on Landfill Project," PT, 28 March 1991.

48. "Mercer Residents Await Signing of Legislation," *Kirksville Daily Express*, 12 May 1991; Suzanne Jones, NPG, 16 May 1991.

49. Executive Order, State of Missouri, May 28, 1991.

50. O'Donnell, "The System and the Environment," p. 5.

51. James F. Wolfe, *NPG*, 30 May 1991.

52. James F. Wolfe, *NPG*, 29 May 1991.

53. James F. Wolfe, "Ashcroft Vetoes Landfill Referendum Bill," *NPG*, 29 May 1991.

54. O'Donnell, "The System and the Environment," p. 5.

55. We are indebted to Janet Eastwood, Grain Millers Union, for material on the union's involvement in the Waste-Tech struggle.

56. On July 10, 1992, Nestlé USA sent the DNR a 42-page report conducted by an engineering firm hired to review the Waste-Tech project application. The report included 257 technical concerns. Nestlé asked the DNR to address these concerns ("Nestlé Issues Report to DNR on Landfill: Asks Department to Stay Review," *PT*, 23 July 1992).

57. One of the commissioners, apparently reluctant to commit herself, had her husband, who was not a commissioner, sign for her.

58. Robert Anderson, "Mercer Panel Rejects Waste-Tech," *NPG*, 29 August 1991.

59. We are indebted to Randi Ferguson for her information regarding the role of New Environmental Winds in the struggle.

60. See, for instance, Elford Horn, *Indians of Mercer County* (Mercer County 1986).

61. We are indebted to Dick Black, official representative of the Iowa Tribe on repatriation in Missouri, for considerable information on the Native American involvement in the Waste-Tech struggle.

62. M. Annette James, ed., *The State of Native America: Genocide, Colonization and Resistance* (Boston: South End Press, 1992).

63. W. Roger Buffalohead, "Reflections on Native American Cultural Rights and Resources," *American Indian Culture and Research Journal* 16, no. 2 (1992), p. 198.

64. An excellent discussion of this trend may be found in "Native North America: The Political Economy of Radioactive Colonialism," in M. Annette Jaimes, *The State of Native America*, pp. 241–66. See also Robert D. Bullard, "Environmental Justice for All: It's the Right Thing to Do," *J. Envtl. Law and Litigation 9*, no. 1 (1994), pp. 301–02. For a useful discussion of Waste-Tech's own targeting of Indian nations, see "Indian Tribes Lured by Money from Toxic Waste Incinerators," *St. Louis Post-Dispatch*, 24 June 1990.

65. Classic studies of environmental racism are Bunyan Bryant and Paul Mohai, *Race and Incidence of Environmental Hazards: A Time for Discourse* (Boulder: Westview Press, 1992); Al Gedicks, *The New Resource Wars*; Bullard and Chavis, eds., *Confronting Environmental Racism*.

66. Sandy Johnson, *Book of Elders* (New York: Harper Collins, 1994).

67. Suzanne Jones, "Indian Burial Ground Cited at Landfill Site," *NPG*, 16 July 1991.

68. "Indian Representative Tells Commission Survey of Waste-Tech Site to Be Requested," *PT*, 18 July 1991.

69. Michael Weichman, state archeologist, refused to supply documentation to Victor Roubidoux, cultural liaison for the Iowa tribe. Though documentation of grave sites is not made public, in order to prevent desecration of grounds by artifact hunters, Roubidoux found it a little surprising that the office of the state archeol-

ogist would imagine that he, an official representing the aboriginal tribe, would be planning to go artifact hunting. Roubidoux believes, instead, that Weichman was acting in the interest of Waste-Tech and did not want official documentation released. It was not until the Carnahan administration that Dick Black, official representative of the Iowa tribe on repatriation in Missouri, was able to obtain the documentation of village site and Native American burial grounds.

70. "Indian Representative Tells Commission."

71. Jim Brantley, "Indian Burial Grounds May Slow Waste-Tech Project," *Trenton Republican Times*, 18 July 1991.

72. Michael Mansur, "Indians Return to Missouri, Fight a Landfill near Ancestors' Graves," *Kansas City Star*, 18 August 1991.

73. Brantley, "Indian Burial Grounds." The CCE January 1992 newsletter commented on the inadequacy of this survey: "The Indians say the archeological study done by the state was designed not to find anything! They have requested the state to do a more indepth study with high-tech sonar equipment in the spring."

74. Robert Anderson, "Indians Hope to Bury Waste-Tech," *NPG*, 18 July 1991. As strict as federal law is, practice in Missouri has been very weak in terms of contacting the aboriginal tribe when Indian remains have been unearthed. In addition, until 1996, with the passage of SB 0834, Missouri law had no section covering trafficking in burial goods. A mound can yield as much as $200,000, as the looting at Slacks Farm, Kentucky, in 1987, proved when several individuals rented ten acres from a farmer and began to dig into the mounds for lucrative artifacts. The Missouri law reads: "A person who commits a crime under this act is guilty of a Class A misdemeanor for the first offense and a Class D felony for a second or subsequent violation."

75. Randi Ferguson and Don Fitz, "Will Amoco Dump Incinerator Ash on Missouri Sacred Indian Lands?" *The Gateway Greens' Compost-Dispatch*, August 1991.

76. Brantley, "Indians Ready to Protect."

77. Mansur, "Indians Return to Missouri."

78. Brantley, "Indians Ready to Protect."

79. Robert Anderson, "Indian Tribes Move In to Protect Burial Mounds," *NPG*, 13 August 1991.

80. Part of this camp was on private land, part on state land. There was, therefore, the matter of state approval. Michael Haney obtained the approval of the Missouri Highway and Transportation Department; the Mercer County Sheriff's Department gave its approval conditioned on the approval of the former.

81. Anderson, "Indian Tribes Move In."

82. Michael Haney, interview by Randi Ferguson, tape recording, Princeton, Mo., Feb. 1993.

83. A lawyer agreed to represent the Iowas pro bono.

84. For the use of the bad-boy law in the Waste-Tech struggle, we are indebted to Ed King, in collaboration with Randi Ferguson, New Environmental Winds activists, for their unpublished "Situation Report" entitled *Amoco/Waste-Tech: The Final Days in Mercer County, MO*.

85. These states are listed in a 1987 Environmental Law Institute Study conducted for EPA.

86. See Brian Lipsett, "Getting Tough on Corporate Crime," *Everyone's Backyard*, June 1992, p. 24, for a discussion of EPA's policies on big/little corporations in terms of debarment.

87. Keven Schafer, ed., (Washington, DC: The Environmental Exchange, 1993), pp. 97–99.

88. Code of State Regulations, chapter 7, Applicable to Owners/Operators of Hazardous Waste Facilities, based on federal regulations 10CSR25-7.270 (2)(H). Though the wording has changed since the dates referred to in the case study, the basic thrust of the law is the same.

89. Ed King's research of the Council of Economic Priorities report, Waste-Tech's Nevada incorporation papers, and many newspaper articles clearly revealed Amoco Oil's ownership of this waste company.

90. Letter dated January 25, 1993.

91. Letter dated January 28, 1993.

92. Cheryl Wittenhauer, "Waste-Tech Landfill Application Clears First Hurdle," NPG, 5 February 1993.

93. Letter dated February 5, 1993.

94. Cheryl Wittenhauer, "State Reviews Waste-Tech Application," NPG, 24 February 1993. Doug Johnson, Waste-Tech's general manager, had admitted that Amoco was the parent company, just "not technically an affiliate." Since the DNR had requested the "ultimate corporate parent," this admission to the press seems perplexing.

95. Michael Mansur, "Landfill Permit Is Withdrawn," The Kansas City Star, 26 February 1993.

96. A news release by Waste-Tech dated February 25, 1993. See also PT, 1 September 1994. Ecova (formerly Waste-Tech) president Tom Noel was quoted as saying, "There is a tremendous oversupply of hazardous-waste incineration capacity, which has led to a substantial decline in prices over the past year." However, Ecova went ahead and built the hazardous waste incinerator at Kimball, Nebraska. According to activists working in the "Stop Dioxin Campaign," the "market forces" claims are bogus. In all cases, incinerators would have been built if the company had not encountered local grassroots opposition. See Lois Gibbs, Dying from Dioxin (Boston: South End Press, 1995), p. 238.

97. Ed King, "The Last Days of Waste-Tech in Mercer County," Compost-Dispatch 4, no. 4 (April 1993), pp. 1, 5–6.

98. According to James G. Trimble, attorney, Waste-Tech's major purpose may have been to handle Amoco's Sugar Creek mess, to "clean up" Amoco's problems in Missouri. For another cover up, though somewhat different, see "Indiana Good Character Law Spurs Attacks on Environmental Agency," The Chicago Environment 1, no. 4 (Spring 1992), pp. 6–7, in which a waste facility attempting to site a facility in Indiana tried to hide its affiliation with a sister company that had violations of federal law for making false statements.

99. Ed King, "The Last Days of Waste-Tech in Mercer County."

100. See, for instance, Ohio's bad-boy law in the case of WTI in Liverpool, Ohio, and Chem Waste's attempt to buy the WTI incinerator in Rachel's Hazardous Waste News #28, 3 June 1992. See also "Regional Recycling Program Shelved," Vancouver Sun, 12 October 1990.

Chapter Five

1. Quoted in Al Gedicks, The New Resource Wars: Native and Environmental Struggles against Multinational Corporations (Boston: South End Press, 1993), p. 67.

2. Gedicks, *The New Resource Wars*, pp. 69, 95, 103. These tactics, largely based on a behavioralist model, sometimes succeed, but often also fail. The literature is full of articles detailing how the people can be hoodwinked into accepting an incinerator, solid or hazardous waste landfill, or other such facility, and much can be learned from looking at these.

3. Sandford Lewis, "The Corporate Right to Cover Up," *Multinational Monitor* 19, no. 5 (May 98), EBSCOhost MasterFile Elite, Item Number 890762.

4. Michael B. Gerrard, *Whose Backyard, Whose Risk* (Cambridge, MA: MIT Press, 1995), p. 86.

5. See Judy Christrup and Robert Schaeffer, "Not in Anyone's Backyard," *Greenpeace* 15, no. 1 (January-February 1990), p. 17; "Wasting the Midwest," *Missouri Ruralist*, 9 September 1989, p. 7, 18; Jim Patrico, "Farmland Going to Waste," *Farm Journal*, November 1989, pp. 24–25.

6. These remarks were made at a public meeting on the proposed waste facility at Princeton, Missouri, on August 9, 1990. Kaufman was emphasizing a trend he observed in rural areas in the Midwest and South. Robert Bullard, one of the more notable researchers on environmental racism, stresses both racism and "class exploitation" as areas needing attention in the "struggle for justice for all Americans." See Robert D. Bullard and Benjamin F. Chavis, Jr., eds., *Confronting Environmental Racism: Voices from the Grassroots* (Boston: South End Press, 1993), p. 206. Looked at from a global perspective, one sees again that both the poorest countries and the poorest parts of developed countries are targeted for waste disposal. Waste companies have not hesitated to dump on white communities that are perceived as politically weak, for example, in England. At the same time, black African countries are under attack.

7. This was tied to entitlement for Superfund money; EPA changed the approach in 1993. See Gerrard, *Whose Backyard, Whose Risk*, p. 50; see also Barry George Rabe, *Beyond NIMBY: Hazardous Waste Siting in Canada and the United States* (Washington, DC: Brookings Institution, 1994), p. 13.

8. See Celeste P. Duffy, "State Hazardous Waste Facility Siting: Easing the Process through Local Cooperation and Preemption," *Boston College Environmental Affairs Law Review* 11(1983–84), p. 770.

9. Lois Gibbs, *Dying from Dioxin* (Boston: South End Press, 1995), p. 278.

10. John Stauber and Sheldon Rampton, *Toxic Sludge Is Good for You* (Monroe, ME: Common Courage Press, 1995), pp. 89–90.

11. Susan Hunter and Kevin M. Leyden, "Beyond NIMBY: Explaining Opposition to Hazardous Waste Facilities," *Policy Studies Journal* 23, no. 4 (1995), p. 601.

12. O'Hare and Sanderson, "Facility Siting and Compensation: Lessons from the Massachusetts Experience," *Journal of Policy Analysis and Management* 12, no. 2 (1993), p. 365.

13. Hunter and Leyden, "Beyond NIMBY," pp. 602, 613. As Hunter and Leyden show, citizens are often motivated by "morality" and "long-standing ideological beliefs," not just by "self-interest."

14. See Douglas J. Lober, "Why Protest? Public Behavioral and Attitudinal Response to Siting a Waste Disposal Facility," *Policy Studies Journal* 23, no. 3 (1995), p. 514.

15. *Edward Abbey: A Voice in the Wilderness*, produced and directed by Eric Temple, 60 min., Eric Temple Productions, 1993, videocassette.

16. Gibbs, pp. 279–82. Argument drawn from Richard Grossman.

17. Sherry Cable and Charles Cable, *Environmental Problems / Grassroots Solutions: The Politics of Grassroots Environmental Conflict* (New York: St. Martin's Press, 1995), p. 109. For typical features of citizens who make up the grassroots environmental movement, see Robert Gottlieb and Helen Ingram, "The New Environmentalists," *The Progressive* (August 1988), pp. 14–15; Christopher J. Bosso, "After the Movement: Environmental Activism in the 1990s," in Norman J. Vig and Michael E. Kraft, eds., *Environmental Policy in the 1990s: Toward a New Agenda* (Washington, DC: CQ Press, 1994), pp. 43–44; Lawrence C. Hamilton, "Concern about Toxic Wastes: Three Demographic Predictors," *Sociological Perspectives* 28, no. 4, (1985), p. 464; Nicholas Freudenberg, *Not in Our Backyards: Community Action for Health and the Environment* (New York: Monthly Review Press, 1984), pp. 133–35. See also Al Gedicks, *The New Resource Wars*, p. 57, on Thomas Gladwin's conclusion that focus has shifted from existing to potential environmental impacts.

18. For this idea, see *Toxic Nation: The Fight to Save Our Communities from Chemical Contamination* (New York: John Wiley & Sons, Inc., 1993), p. 108.

19. *The Backyard Revolution: Understanding the New Citizens' Movement* (Philadelphia: Temple University Press, 1980), pp. 1–43.

20. Stella M. Capek, "The 'Environmental Justice' Frame: A Conceptual Discussion and an Application," *Social Problems* 40, no. 1 (1993) p. 9.

21. For this emphasis on equal partnership in building a just society at the community level, see Richard Hofrichter, ed., "Introduction," *Toxic Struggles* (Philadelphia: New Society Publishers, 1993), p. 2; Lois Gibbs, *Dying from Dioxin*; Stephen Horton, "Value, Waste, and the Built Environment: A Marxian Analysis," *Capitalism, Nature, Socialism* 8, no. 1 (March 1997), pp. 127–39.

22. Cable and Cable, *Environmental Problems*, p. 112.

23. Gibbs, *Dying from Dioxin*, chap. 16.

24. Freudenberg, *Not in Our Backyards*, p. 172.

25. Ibid., p. 144.

26. Gedicks, *The New Resource Wars*, p. 111.

27. Supra, chap. 4. 56.

28. See Judy Christrup and Robert Schaeffer, "Not in Anyone's Backyard," *Greenpeace* 15, no. 1 (January-February), p. 17, indicating that "state politicians are pushing for legislation to do away with township zoning."

29. See Harold A. Ellis, "Neighborhood Opposition and the Permissible Purposes of Zoning," *J. Land Use & Envtl. L*, 7 (1992).

30. Thomas E. Cronin, *Direct Democracy: The Politics of Initiative and Recall* (Cambridge, MA: Harvard University Press, 1989), p. 52; V. O. Key, Jr., and Winston W. Crouch, *The Initiative and the Referendum in California* (Berkeley: University of California Press, 1938), p. 437.

31. Key and Crouch, *The Initiative and the Referendum in California*, p. 438.

32. Ellis Paxson Oberholtzer, *The Referendum in America* (New York: Charles Scribners Sons, 1912), p. 208.

33. Ibid., pp. 209, 311.

34. Ibid., p. 319.

35. Ibid.

36. *Encyclopedia Britannica*, Chicago, 1973, s.v. Referendum. See *Pacific Telephone and*

Telegraph Co. v. Oregon, 233 U.S. 118 (1912). It has been observed by Carl Friedrich that "the electorate is representative of the whole people."

37. Cronin, *Direct Democracy*, p. 47.

38. Ibid., p. 197.

39. Ibid., pp. 199–200.

40. Ibid., p. 203.

41. Ibid., pp. 219–20.

42. Key and Crouch, *The Initiative and the Referendum in California*, pp. 504-05.

43. Ibid., p. 574.

44. Hugh F. O'Donnell, III, "The System and the Environment: Another Example of the Erosion of Democracy," unpublished manuscript, pp. 5–6.

45. A second nonbinding referendum, Senate Bill 55, was also introduced by Senator Danner and passed by the full Senate on February 24, 1993, prior to the Waste-Tech pullout. Binding referendums, giving the voters a say in the permitting process, were also introduced by both Danner and Tate. Danner's Senate version, Senate Bill 267, was much amended and stalled in committee; Tate's was heard in committee on the day of the Waste-Tech pullout. In 1993, Tate also was successful in passing legislation that prohibited the permitting of a hazardous waste facility in a non-karst area of the state over a groundwater divide (Section 260.429 RSMo). This category applied to north Missouri, Tate told the press (*Trenton Republican Times*, 19 May 1993).

46. We are indebted to Dick Black here and elsewhere in this chapter for his assistance in Native American legal, cultural, and religious issues.

47. The United States Indian Court of Claims determined aboriginal title for Iowas in 1955.

48. Quoted in *What Works Report No. 2: Local Solutions to Toxic Pollution* (Washington, D.C.: The Environmental Exchange, 1993), p. 98.

49. See "Indians Are Not Specimens—Indians Are People," *Association on American Indian Affairs* (Fall 1989), pp. 3–4; "A History of the American Indian Religious Freedom Act and Its Implementation," *Indian Affairs* (Summer 1988), p. ii.

50. Jack F. Trope and Walter R. Echo-Hawk, "The Native American Graves Protection and Repatriation Act: Background and Legislative History," *Arizona State University Law Journal* 24, no. 1 (Spring 1992), p. 38. For a comprehensive discussion of the differences between White and Indian attitudes regarding the issue of sanctity toward burials, see Randall H. McGuire, "The Sanctity of the Grave: White Concepts and American Indian Burials," in Robert Layton, ed., *Conflict in the Archeology of Living Traditions* (London: U. Hyman, 1989), pp. 167–84. For the issue of diversity in Native American beliefs and practices regarding burials, see Paul G. Bahn and P. W. K Paterson, "The Last Rights: More on Archeology and the Dead," *Oxford Journal of Archeology* 5, no. 3 (1986), p. 261. But see also Ronald L. Grimes, "Desecration of the Dead: An Inter-religious Controversy," *American Indian Quarterly* 10 (Fall 1986), p. 309, who claims a Native American consensus on the issue of cemetery desecration.

51. Guest Essay, *Native Peoples Magazine*, Spring 1994. The belief that the earth might be viewed "not only as a burial ground but as a 'church'" seems relevant here (quoted in Grimes, "Desecration of the Dead," 307.) See also McGuire, "The Sanctity of the Grave," who states, "The concept of ancestry [Indians] apply to the dead is a communal one that requires respect for the sanctity of the grave even in the absence of direct familial relations" (p. 180).

52. Kathryn Milun, "(En)countering Imperialist Nostalgia: The Indian Reburial Issue," *Discourse* 14, no. 1 (Winter 1991-92), p. 67.

53. See, for instance, Trope and Echo-Hawk, "The Native American Graves Protection and Repatriation Act," pp. 38–45; June Camille Bush Raines, "One Is Missing: Native American Graves Protection and Repatriation Act: An Overview and Analysis," *American Indian Law Review* 17, no. 2 (Summer 1992), pp. 642–44; Milun, "(En)countering Imperialist Nostalgia," pp. 62–64.

54. Michael Haney commented, "Where we come from this is not a landfill issue, it is a human rights issue" (Brantley, "Indian Burial Grounds"). For excellent articles providing insight into NAGPRA as human rights legislation, see Raines, "One Is Missing," 639–64; Trope and Echo-Hawk, "The Native American Graves Protection and Repatriation Act," pp. 35–77.

55. Grimes, "Desecration of the Dead," pp. 305–18.

56. Videotape by New Environmental Winds.

57. Laurie Anne Whitt, "Cultural Imperialism and the Marketing of Native America," *American Indian Culture and Research Journal* 19, no. 3 (1995), p. 131. As examples, Whitt mentions pharmaceutical and musical piracy.

58. Whitt's statement that human skeletal remains "found on federal public lands . . . belong to the U.S. government, provided they are at least one hundred years of age" leaves out the whole issue of tribal ownership and control under PL 101-601, Sec. 3, an issue of vital importance to this chapter. For articles dealing with the government's historic protection of "archeological resources" on federal land, see "No Peace for Indian Burial Grounds," *Indian Affairs* (Summer 1988), p. ix; Raines, "One Is Missing," pp. 649–55; Trope and Echo-Hawk, "The Native American Graves Protection and Repatriation Act," pp. 40–43.

59. See, for instance, C. Dean Higginbotham, "Native American Versus Archaeologists: The Legal Issues," *American Indian Law Review* 10, no. 1 (1982), pp. 91–115; John E. Peterson II, "A Conflict of Values: Legal and Ethical Treatment of American Indian Remains," *Death Studies* 14, no. 6 (1990), pp. 519–54.

60. Peterson, "A Conflict of Values," p. 525.

61. Bahn and Paterson, "The Last Rights: More on Archeology and the Dead," p. 261. See Milun, "(En)countering Imperialist Nostalgia," p. 69, who states that, in fact, the scientific community "actually admits that it does not have a way of extracting the information it claims these bones contain"; see also World Council of Indigenous Peoples, "The Sacred and the Profane: The Reburial Issue as an Issue," *Death Studies* 14, no. 6 (1990), pp. 504–5.

62. See Bahn and Paterson, "The Last Rights," pp. 261–68. The way to balance these claims is to develop a system of compromises. Bahn and Paterson identified the best solution for scholars to be "acquiescence, compromise, and collaboration." One alternative would be "easy access" to protect against further decomposition. A second alternative would be "a system of 'keeping places' . . . where material is in the care of Native Americans but accessible, on application, to serious scholars." See also Peterson, "A Conflict of Values," p. 522, on physical anthropologists' reluctance to see reburial.

63. See, for instance, McGuire, "The Sanctity of the Grave." See also Peterson, "A Conflict of Values," pp. 519–54. Peterson takes up an interesting question: Is the issue, then, the traditional conflict between science and religion? One argument is that the conflict "is not a science vs. religion controversy but a clash between conflicting religions." Instead of "two different kinds of systems," religion and science,

the clash is "between competing value systems." The "religion" of the archeologist is "Western humanism"; the religion of the Native Americans, allowing for differences in tribal beliefs, is "North American intertribalism." The claims of scholarly humanists, like religious claims in general, "assume the universality" of these scholars' codes of ethics. Besides this claim of universality, scholars also assume that their research is "neutral." But from the Native American point of view, as Grimes points out in "Desecration of the Dead," pp. 309-13, the "mode of acquisition is an affront to native religious values."

64. Lois Gibbs, *Dying from Dioxin*, p. 238.

65. E-mail correspondence with Bill Geary, Corporate Public Affairs, Clean Harbors, Inc., Braintree, Mass. Clean Harbors is also a company that provides hazardous waste services for the U.S. government (www.cleanharbors.com.). Clean Harbors is a NASDAQ company. Nine-month revenues for the company for 1997 were $138 million, with a net loss claimed of $3.6 million. The company claims to be a "major American environmental services provider" with 12,000 customers, including 300 of the *Fortune* 500 companies. The company has 11 hazardous waste facilities in the U.S. Kimball was the company's only incinerator in 1999. According to Geary, "Ecova was created by Amoco for the sole purpose of constructing the facility and thereafter operating it. The facility was initially intended to incinerate oil soaked soil and other contaminated materials generated from various Amoco oil refineries. Ecova also intended to market the incinerator's excess disposal capacity beyond their own needs to American industry." It took eight years for the facility to be permitted and constructed. The facility passed the trial burn testing in December 1994. Once sold to Clean Harbors, Ecova Corp. was dissolved. It is capable of processing solids, sludges, slurries, and liquid wastes at a maximum rate of 8.5 tons per hour. This facility is known as a bubbling fluid-bed combustor. The bed is 7.5 feet by 15 feet with a 60-foot stack and a rating of 40 million BTU per hour heat release. It uses scrubbers and a fabric-filter bag house for dust removal. According to Geary, the facility operated for some of the five years prior to 1999 at near capacity. For operating parameters of bubbling-bed hazardous waste combustors and schematic depictions, see Carl H. Fromm and Charles D. Bartholomew, "Fluidized-Bed Incineration," in Harry M. Freeman, ed., *Standard Handbook of Hazardous Waste Treatment and Disposal* (New York: McGraw-Hill, 1998), pp. 8.33–8.40.

66. Testimony of Linda Briscoe and Rev. Solomon Lundy, Co-Chairs of the Toxics/Pollution Committee of Committees United for Action, before the Committee on Environment and Public Works of the United States www.environlaw.com/cepw/ index.html.

67. "EPA Renews Attack on Kentucky Audit Privilege and Immunity Law," *Chem Trends, a publication of the Associated Industries of Kentucky Chemical Industry Council* 8, no. 1 (First Quarter, 2001), www.aik.org/cic/c+1q2001/chemtrends% 201st%20quarter.

68. Philip Weinberg, " 'If It Ain't Broke...': We Don't Need Another Privileges and Immunities Clause for Environmental Audits," *Journal of Corporation Law* 22, no. 3 (Spring 97), EBSCOhost MasterFILE, Item Number 9707162986.

69. "Fort Wayne Citizens Winning in Fight Against WMI," www.stopwmx. org/ bad-boy.html.

70. See especially "Absolution for Polluters: Easy Forgiveness for Industry's Dirty Secrets," *Sierra* 80, no. 4 (July-August 95), EBSCOhost, MasterFILE, Item Number

9508102894. See also Weinberg, "'If It Ain't Broke.'"; "The Privileged Class: Bush Pushes Secrecy for Environmental Audits," www.txpeer.org/Bush/privileged_class.html.

71. Yet, according to the EPA, no civil or criminal penalties are imposed based on voluntary audits. See, for instance, Charmaine Oakley, "Guilt-Free Eco-Audits," *Earth Island Journal* 12, no. 1 (Winter 96/97), EBSCOhost MasterFILE, Item Number 9702042177.

72. Peter Montague, "Right to Know Nothing," *Rachel's Environment and Health Weekly*, # 552, June 26, 1997, www.monitor.net/rachel/r552.html; see also Weinberg, "'If It Ain't Broke.'"

73. Montague, "Right to Know Nothing."

74. Most recently, HB786, with last action February 20, 2001, referred to Judiciary. Hearing not scheduled. Bill currently not on calendar.

75. Quoted in "Absolution for Polluters."

Chapter Six

1. Juan Martinez-Alier, "Environmental Justice (Local and Global)," *Capitalism, Nature, Socialism* 8, no. 1 (March 1997), p. 104.

2. Slogan of Waste Management, Inc. For an interesting counter to such a sentiment, see Stephen Horton, "Value, Waste and the Built Environment: A Marxian Analysis," *Capitalism, Nature, Socialism* 8, no. 1 (March 1997), p. 139.

3. Horton, "Value, Waste and the Built Environment," p. 139.

4. Quoted in Horton, p. 139.

5. Robert Bullard, *Confronting Environmental Racism: Voices from the Grassroots* (Boston: South End Press, 1993), pp. 19–20.

6. Mutombo Mpanya, "The Dumping of Toxic Waste in African Countries: A Case of Poverty and Racism," in Bunyan Bryant and Paul Mohai, eds., *Race and the Incidence of Environmental Hazards* (Boulder, CO: Westview Press, 1992), pp. 204–14.

7. William Lash III, "Environment and Global Trade," *Society* 31, no. 4 (May–June 1994), p. 54.

8. *Waste Not*, #360, March 1996.

9. Ibid.

10. Center for Investigative Reporting and Bill Moyers, *Global Dumping Ground: The International Traffic in Hazardous Waste* (Washington: Seven Locks Press, 1990); Eddie J. Girdner, *People and Power* (Istanbul: Literatur, 1999), pp. 396–97.

11. Jennifer Clapp, "The Toxic Waste Trade with Less-Industrial Countries: Economic Linkages and Political alliances," *Third World Quarterly* 15, no. 3 (1994), pp. 510-12; Mutombo Mpanya, "The Dumping of Toxic Waste in African Countries: A Case of Poverty and Racism," in Bryant and Mohai, eds., *Race and the Incidence of Environmental Hazards*," pp. 204–14; Gareth Porter and Janet Welsh Brown, *Global Environmental Politics* (Boulder: Westview Press, 1991), pp. 86–87.

12. John T. Aquino, "ENTSORGA Speaks: Confusing Definitions from Basel Convention Hurt Recycling," *Recycling Times*, no date, www.wasteage.com.

13. Clapp, "Toxic Waste Trade," pp. 509–13. A new agreement was signed in 2000. See "Europa/Development - New ACP-EU Agreement," http://europa.eu. int/comm/development/ cotonou/index_en.htm.

14. U.S. Environmental Protection Agency, Office of Pollution Prevention and Toxics, *1996 Toxics Release Inventory* (Washington, DC: U.S. Government Printing Office, 1998), pp. 16–17.

15. Ibid.

16. Lash, p. 55.

17. Ibid., p. 57.

18. Robert Weissman, "Corporate Plundering of Third-World Resources," in Hofrichter, ed., *Toxic Struggles*, pp. 186–96. Another case is seen in Cyprus, victim of both the American mining company, Cyprus Amax Mining Corporation, and the state. Eddie J. Girdner, "Aphrodite's Nightmare: Cyprus Mining Company and Environmental Disaster in Northern Cyprus," *Scandinavian Journal of Development Alternatives and Area Studies* 18, nos. 2 & 3 (June-September 1999).

19. E. F. Schumacher, *Small Is Beautiful: A Study of Economics as if People Mattered* (New York: Perennial Library, 1975), p. 44.

20. Subir Sinha and Ronald Herring, "Common Property, Collective Action and Ecology," *Economic and Political Weekly* 28, nos. 27 & 28 (July 3–10, 1993), pp. 1425–32.

21. Paul Ekins, *A New World Order: Grassroots Movements for Global Change* (London: Routledge, 1992), pp. 38–39.

22. Claire Van Zevern, "Land, Ecology, and Women: Implications For Hawaiian Sovereignty," in Jennifer Turpin and Lois Ann Lorentzen, *The Gendered New World Order* (New York: Routledge, 1996), pp. 83–94; "A land ethic...reflects the existence of an ecological conscience . . .": Aldo Leopold, *A Sand County Almanac* (New York: Ballantine Books, 1970), p. 238.

23. Van Zevern, p. 88.

24. Ibid., p. 89.

25. Schumacher, *Small Is Beautiful*, p. 20.

26. Ibid.

27. Ibid., p. 23.

28. Ekins, *A New World Order*, pp. 204–05.

29. Debaranjan Sarangi, "People's Resistance to TISCO Project," *Economic and Political Weekly* 31, no.13 (March 30, 1996), pp. 809–10.

30. "Netarhat Project: Biggest Ever Tribal Displacement," *Economic and Political Weekly* 29, no. 18 (April 30, 1994), pp. 1055–56.

31. Stephen Rego, "Destructive Development and People's Struggles in Bastar," *Economic and Political Weekly* 29, no. 7 (February 12, 1994), pp. 352–53.

32. Ekins, *A New World Order*, pp. 88–94.

33. Ibid., p. 93.

34. "Sardar Sarovar Project: Review of Resettlement and Rehabilitation in Maharashtra," *Economic and Political Weekly* 28, no. 34 (August 21, 1993), pp. 1705–14.

35. Philip D'Souza, et. al., "Bargi Dam: Oustees Pay Price of Development," *Economic and Political Weekly* 31, nos. 45 & 46 (Nov. 9–16), pp. 2984-85.

36. Jashbhai Patel, "Is National Interest Being Served by Narmada Project?" *Economic and Political Weekly* 29, no. 30 (July 23, 1994), pp. 1957–64. In India, for example, some 20 million people have been uprooted from their ancestral villages and communities since 1947 by new dams.

37. Krishna Kumar, "State and People's Styles of Suppression and Resistance," *Economic and Political Weekly* 31, no. 39 (September 28, 1996), pp. 2666–67.

38. Sanjay Sangvai, "Re-opening Sardar Sarovar Issue," *Economic and Political Weekly* 30, no. 11 (March 18, 1995), pp. 542–44.

39. Sanjay Sangvai, "CM's Meeting on Narmada: What Did Not Happen," *Economic and Political Weekly* 31, no. 34 (August 24–31, 1996), pp. 2287–88.

40. V. R. Krishna Iyer, "Nature Friendly Planning of Humanity's Future: Dialectics and Dynamics of Development Management," *Economic and Political Weekly* 31, no. 34 (August 24–31, 1996), pp. 2297–2300.

41. Krishna Kumar, "State and People," p. 2667.

42. Dunu Roy, "Large Projects: For Whose Benefit?" *Economic and Political Weekly* 29, no. 50 (December 10, 1994), p. 3129.

43. Ibid.

44. Rajni Bakshi, "Development, Not Destruction: Alternative Politics in the Making," *Economic and Political Weekly* 31, no. 5 (February 3, 1996), p. 255.

45. Ibid., pp. 256–57. The final decision of the Supreme Court of India on the Narmada Dam (*Narmada Bachao Andolan v. Union of India*, October 10, 2000) ruled that the project could go ahead. The Supreme Court ruled that displaced people have the "right to live in a healthy environment, right to a house," and the "right to food," but that displaced people, especially tribal Adivasis, are "being actually helped by the state since their displacement will provide an opportunity to uplift them from their present living conditions." See Philippe Cullet, "Sardar Sarovar Judgement and Human Rights," *Economic and Political Weekly* 36, no. 18 (May 5, 2001), p. 1503. See also Shiv Visvanathan, "Supreme Court Constructs a Dam," *Economic and Political Weekly* 35, no. 48 (November 25, 2000), pp. 4176–80, where it is noted that the Supreme Court ruled that for the displaced tribal persons, "the gradual assimilation in the mainstream of society will lead to betterment and progress" (p. 4179). The author notes that "[t]here is not much wrong with the court's idea of law but it is an idea of law imbedded in a philosophy of progress, development and change that would make any Ford Foundation/ USAID/IDS expert of the sixties happy" (p. 4177). In spite of the decision of the Indian Supreme Court, the struggle for justice for those displaced by large dams has refocused the issue on rehabilitation and the right of the uprooted people to participate in this process. See Anant Phadke, "Dam-Oustees' Movement in South Maharashtra," *Economic and Political Weekly* 35, no. 47 (November 18, 2000), pp. 4084–86. In some cases governments are being forced to cancel large dams for small-scale watershed projects that avoid displacement altogether.

46. Ekins, *A New World Order*.

47. Ibid., pp. 78–80.

48. Ibid., pp. 81–84.

49. Ibid., pp. 100–111.

50. Binay Pattnaik and Anirudh Brahmachari, "Community-Based Forest Management Practices," *Economic and Political Weekly* 31, no. 15 (April 13, 1996), pp. 968–75.

51. Makal, "Villages of Chipko," *Economic and Political Weekly* 28, no. 15 (April 10, 1993), pp. 617-21; Ekins, *A New World Order*, pp. 143–44.

52. Makal, "Villages of Chipko."

53. Pattnaik and Brahmachari, "Community-Based Forest Management Practices," pp. 970–72.

54. Ekins, *A New World Order*, pp. 81–86.

55. Amita Baviskar, "Fate of the Forest: Conservation and Tribal Rights," *Economic and Political Weekly* 29, no. 38 (September 1994), pp. 2473–2501.

56. Ibid., p. 2500.

57. Sinha and Herring, "Common Property, Collective Action and Ecology."

58. Ibid.

59. Ibid.

60. Suman Sahai, "How Do We Protect Our Genetic Resources?" *Economic and Political Weekly* 31, no. 27 (July 6, 1996), p. 1725; Vandana Shiva, "Agricultural Biodiversity, Intellectual Property Rights and Farmers' Rights," *Economic and Political Weekly* 31, no. 25 (June 22, 1996), p. 1623.

61. Vandana Shiva, "Agricultural Biodiversity, Intellectual Property Rights and Farmers' Rights," p. 1629.

62. K. Ravi Srinvas, "Sustainable Agriculture, Biotechnology and Emerging Global Trade Regime," *Economic and Political Weekly* 31, no. 29 (July 20, 1996), pp. 1922–23.

63. Ghayur Alam, "Biotechnology, Agriculture and Developing Countries," *Economic and Political Weekly* 31, no. 12 (March 23, 1996), pp. 703–04; M. G. G. Pillai, "Multinationals and the Environment," *Economic and Political Weekly* 31, no. 6 (February 10, 1996), p. 325.

64. Vasant Kumar Bawa, "Gandhi in the 21st Century: Search for an Alternative Development Model," *Economic and Political Weekly* 31, no. 47 (November 23, 1996), pp. 3048–49.

65. Princen and Finger, *Environmental NGOs in World Politics*, p. 225.

Index